Advances in Seafood Biochemistry: Composition and Quality

Advances in Seafood Biochemistry: Composition and Quality

Arnold C. Long

CYBER TECH PUBLICATIONS
4264/3, Ansari Road, Daryaganj, New Delhi-110002 (India)
Ph.: 011-23244078. Fax: 011-23280028
E-Mail: cyberpublicationsdelhi@yahoo.com

ADVANCES IN SEAFOOD BIOCHEMISTRY: COMPOSITION AND QUALITY

Arnold C. Long

First Edition 2009

Published by :

G.S. Rawat for **Cyber Tech Publications**
4264/3, Ansari Road, Daryaganj, New Delhi-110002 (India)
Ph.: 011-23244078. Fax: 011-23280028
E-Mail: cyberpublicationsdelhi@yahoo.com

Printed at : **Amit Enterprises**, Maujpur, Delhi, (9810878184)

Preface

Doctors have recommended including seafood as part of a healthy diet because of the variety of nutrients it provides. However, contamination of seafood, whether by naturally occurring or introduced contaminants, remains a concern for consumers because of the potential for adverse health effects. The extent to which a contaminant in a food may be considered a risk to health depends upon the nature and level of the compound present, and the sensitivity of individuals or groups in a population to potentially toxic compounds. Specific population groups have been identified as being at particular risk from exposure to contaminants in seafood. Paradoxically, these population groups may especially benefit from the nutrients in seafood. For most of the general population, optimal benefits from seafood can be obtained by making choices to maximize intake of desirable nutrients balanced against exposure to contaminants that may pose a health risk. Making such selections, however, requires that consumers are aware of the variety of seafood available and are provided information on both benefits and risks to inform their choices.

For the purposes of this book, the term seafood refers to all commercially obtained fish, shellfish, and mollusks, both marine and freshwater. When marine mammals are pertinent to the discussion, they will be identified separately. The impact of seafood obtained by subsistence and recreational harvesting is considered as far as the more limited data allow.

—*Author*

CONTENTS

Preface

1. Chemical Contaminants in Seafood 1
2. Health Risk Assessment 24
3. Nature of Public Health Risks 106
4. Aquatic Genetic Resources 130
5. Naturally Occurring Fish and Shellfish Poisons 166
6. Production, Distribution, and Consumption 195
7. Microbiological and Parasitic Exposure 207

Bibliography 277

Index 279

1

CHEMICAL CONTAMINANTS IN SEAFOOD

This chapter and the following one should be considered as a unit. Although the committee has not attempted a comprehensive quantitative assessment of the risks of chemical contaminants in seafood, this chapter performs the functions of the "hazard identification" portion of a chemical risk assessment–giving a broad overview of many different potential seafood contamination problems, as well as an extensive summary of available data for characterizing contaminant concentrations in aquatic organisms in the environment.

Next Chapter provides a discussion of the directions needed to improve quantitative risk assessment in this area [including a detailed treatment of the methods used to assess two specific hazards (polychlorinated biphenyls and methylmercury)] and, more broadly, covers the issues that are usually found under the headings "dose-response assessment," "exposure assessment," and "risk characterization," as well as some risk management considerations.

The inorganic contaminants with the greatest potential for toxicity appear to be antimony, arsenic, cadmium, lead, mercury, selenium, and sulfites (used in shrimp processing). Among organic compounds, polychlorinated biphenyls, dioxins, several chlorinated hydrocarbon insecticides, certain processing-related contaminants (nitrosamines and possibly products of chlorination), and contaminants related to aquaculture pose sufficient potential risks for consumers to be worthy of additional study. In addition to providing a broad survey of data on chemical contamination of aquatic organisms and potential risks, this chapter undertakes an extensive set of analyses of the variability of concentrations of certain contaminants across geographic areas and the implications of this variability for control.

In general, lognormal distributions appear to provide good descriptions of the pattern of variation of chemical contaminant concentrations among different geographic areas, and some contaminants (mostly organics) appear to be much more variable than others. The variability of contaminant concentrations among geographic areas is important because it indicates the potential for reduction of exposure through restrictions on the harvesting of aquatic organisms from specific sites. Based on analyses of data for inshore marine waters, for the most variable contaminants/sets of species, it would be possible theoretically to reduce the population dosage delivered by more than 50% by restricting harvesting/marketing from only the 5% most intensely contaminated sites. There is, therefore, considerable potential for management of the overall population dosage of contaminants by measures that would restrict harvesting in specific ways.

INTRODUCTION

There is no area of the committee's work that poses greater challenges to both the scientific tools for understanding likely health hazards and the social tools for managing risks, than the diverse collection of chemical residues that find their way into the human diet partly by way of seafood. Moreover, the confusion between technical questions and social control problems is connected. The understanding of toxicology and environmental health has made important strides since the multitiered structure of federal food protection law was erected (principally by legislation in 1906, 1938, and 1958). Older concepts, which shaped the legislative framework within which food protection agencies attempt to function, suggested sharp distinctions between "poisons" and other substances, or between "safe" and "unsafe" levels of exposure to important categories of environmental toxicants.[1] These ideas are gradually giving way to a more quantitative (although generally still highly uncertain) conception of risks, based on more detailed information about the mechanisms by which different substances interact with intricate biological systems and the diversity of those systems in different individuals in the large and disparate human population. To the extent that increased understanding indicates that certain categories of risks cannot be eliminated entirely, the tools for social control of these risks will have to be adapted to manage toxicant exposures and risks in the light of explicitly formulated trade-offs between the costs of forgoing certain portions of our food resources and the costs of potential adverse effects.

The technical advances that have occurred in risk assessment in recent years have been applied most readily to issues of health protection by governmental institutions of relatively late vintage, operating under legislation adopted within the last 20 years – most notably, the different branches of the U.S. Environmental Protection Agency (EPA) and analogous state authorities. When the more modern techniques and assumptions for quantitatively assessing risks are applied to seafood contaminants, there are a number of areas of mismatch[2] that give the appearance of inconsistency in the social and technical judgments on risks made by different agencies.

Both this chapter and the next deal with aspects of chemical residues in aquatic organisms. In this chapter the committee focuses on the tasks that are usually thought of as part of the hazard identification portion of a quantitative chemical health risk assessment.

The committee has not, however, attempted a formal and comprehensive assessment of the risks of chemical residues in aquatic organisms. Aside from the fact that the available data on both contaminant[3] levels and risks are inadequate for such a task at this time, the charge to the committee emphasized the review of the adequacy of current risk recognition, risk assessment, and risk management procedures in governmental agencies. In the next section of this chapter, the committee gives a broad overview of the types of toxic agents that are known or believed to be contaminants of seafood. Then, the various data bases available for characterizing the geographic and species distribution of chemical contaminants are reviewed, followed by the quantitative insights gleaned from these sources. The variability of contaminants by geography and species, which provides some of the most potentially important opportunities for reduction of exposures, is then considered. Finally, preliminary conclusions and recommendations are offered, based on the material discussed in this chapter.

TOXIC AGENTS AND POTENTIAL TYPES OF HEALTH EFFECTS

Metals and Other Inorganics

The human and veterinary medical literature is replete with information regarding the toxicity of heavy metals. Based on this information, different metals can be classified as having major, modest, minor, or no potential for toxicity. Those with major potential for toxicity, in the committee's view, are antimony, arsenic, cadmium, chromium, lead, mercury, and nickel. Contaminants with a modest potential for toxicity include copper, iron, manganese, selenium, and zinc. Those of minor or no toxicity are aluminum, silver, strontium, thallium, and tin. This classification is based, among other parameters, on potency for producing effects and accessibility of the toxicant. Thus, such metals as nickel and chromium, known inhalant carcinogens, are among those of greatest toxicity, whereas selenium and tin are placed in the lesser categories. When considering the same metals as contaminants of an aquatic food source, however, their relative toxicities will certainly change. Criteria for identifying contaminants (hazard assessment, hazard analysis) of public health concern in the aquatic environment may vary but have been defined (PTI, 1987). These include (1) persistence, (2) bioconcentration potential, (3) toxicity to humans (or suspected toxicity), (4) sources of contaminants in the area of interest, and (5) high concentration in fish and shellfish from the area of interest.

By applying such criteria, both nickel (except for its carbonyl form) and chromium (at least in its hexavalent form), inhalant carcinogens and elicitors of dermal hypersensitivity, would be suspect as contaminants of public health concern (Haines and

Nieboer, 1988). However, both are poorly absorbed from the gastrointestinal tract, and there is little evidence that this route of exposure results in systemic toxicity (Beliles, 1978; Nieboer and Jusys, 1988). Similarly, the use of tri*n*-butyltin (TBT) to control marine fouling of vessels and aquaculture sea pens has been followed by the accumulation of butyl- and elemental tins in the muscles of fish and invertebrates (Short and Thrower, 1987a,b).

Organic tin compounds tend to be more toxic than inorganic salts, and organic forms in particular may be of public health concern. Although little information exists about the toxicity of tin to man, there is sufficient information regarding dosage levels without observable effect to eliminate the probability of tin poisoning from contaminated seafood (WHO, 1980). Conversely, selenium is well recognized as toxic by ingestion and, at existing levels in some seafoods, may be a source of risk (Fan et al., 1988). Antimony has been recognized as both an occupational and an iatrogenic toxicant (Anonymous, 1988a,b; Groth et al., 1986). Recent seafood residue studies, however, either have failed to sample for this metal or indicate concentrations above detectable levels in few contaminated sites (Lowe et al., 1985; NOAA, 1987). Such findings and reasoning, coupled with estimates of ingestion levels, suggest a preliminary list of heavy-metal contaminants found in the edible portions of aquatic animals that may be detrimental to human health. The metals identified in this hazard analysis include arsenic, cadmium, lead, mercury, and selenium. Some toxicity information related to antimony is given later, in the section where current dosage is compared to "acceptable daily intake" levels and other recommended standards.

SPECIFIC TRACE METALS

Arsenic

Arsenic has a long history as a potent poison of humans and other animals. Previously used as a chemotherapeutic and homicidal drug, much information has been collected regarding its toxicity. It exists as the toxic trivalent form (arsenic trioxide, sodium arsenate, arsenic trichloride, etc.), as the less toxic pentavalent form (arsenic pentoxide, arsenic acid, lead arsenate, calcium arsenate, etc.), and as numerous organic forms (arsanilic acid, bimethyl arsenate, etc.). When ingested, inorganic arsenic may cause acute or chronic toxicity and is of primary concern as a carcinogen responsible for pulmonary carcinoma, hemangiosarcomas, and dermal basal cell and squamous cell carcinomas.

Its toxicity is dependent on oxidation state and route of exposure. In its chronic manifestations, arsenic is responsible for gastroenteritis, nephritis, hepatomegaly, peripheral symmetrical neuropathy, and a number of lesions of the skin including plantar and palmar hyperkeratosis and generalized melanosis.

Some of these lesions appear related to destruction of capillary endothelium, with consequent edema and even circulatory failure. At the molecular level the metal is known to uncouple phosphorylation; to react with sulfhydryl groups, thus upsetting cellular metabolism; to damage deoxyribonucleic acid (DNA) directly and to inhibit its repair (Buck, 1978). In addition, as sodium arsenate and arsenite it is teratogenic in lower animals (Earl and Vish, 1978). The metal, therefore, places at special risk pregnant and nursing mothers and their children.

However, the predominant form of arsenic that exists in the edible portions of aquatic animals is the organic form, either arsenobetaine or arsenocholine. These forms have been named "fish arsenic" and no toxic effects from their ingestion have been reported in animals [at doses of 10,000 milligrams (mg) per kilogram (kg)] or in humans.

Furthermore, there is no evidence of mutagenicity by arsenobetaine (Penrose, 1975; Tam et al., 1982). Although arsenobetaine constitutes the bulk of arsenic in fish, available studies are inadequate to conclude that the amounts of more toxic inorganic forms of arsenic (or organic forms that can be metabolized to inorganic arsenic in humans) are negligible in all fish. It is known, however, that the trivalent form (inorganic) is toxic to man and that long-term effects include dermal hyperkeratosis, dermal melanosis and carcinoma, hepatomegaly, peripheral neuropathy, and in cases of inhalation, pulmonary carcinoma (ATSDR, 1989a; Goyer, 1986).

Arsenic is used in the manufacture of pesticides, herbicides, and other agricultural products and is a by-product of mining and smelting operations (Buck, 1978).

Cadmium

Cadmium is unique among toxic metals because it is a relatively recent (50 years) contaminant of the aquatic environment. Its sources are solid waste dumping (pigment in paint) and cadmium-containing sewage sludge, the use of phosphatic fertilizers, electroplating and galvanizing manufacture, and mining (zinc, lead) wastewater (Sherlock, 1986; Sloan and Karcher, 1985). Cadmium is commonly found in its metallic form and as sulfides and sulfates. Invertebrates, both crustacea and bivalves, tend to accumulate metallic cadmium in large amounts by binding to various high-molecular-weight metallothioneine ligands. There is a differential affinity between crustacean muscle and hepatopancreas, the latter organ containing 10-20 times the concentration of the former. Because hepatopancreas may be considered a delicacy or marketed as "brown crab meat," the potential for ingesting large amounts of cadmium when eating lobsters or crabs is increased (McKenzie-Parnell et al., 1988; Sloan and Karcher, 1985).

Cadmium may damage cells by its activity in the plasmalemma where it reacts with phosphate groups of the lipid bilayer to alter permeability, in the nucleus where it is mutagenic, on lysosomal membranes, and as an inhibitor of mitochondrial activity. Its ability to stimulate metallothioneine production in aquatic animals, however, does much to decrease its toxicity (Viarengo, 1985).

Cadmium has been responsible for major human poisoning incidents as a contaminant of wastewater used for irrigation in Japan where the illness is known as itai-itai (ouch-ouch) disease. It is a chronic osteoporotic and osteomalacic condition that primarily affects multiparous females (Kobayashi, 1978). Although the highest accumulation of cadmium is found in bone, the liver and kidney also have a propensity for accumulating the metal, and the kidney is often seriously damaged in chronic occupational exposures (Lauwerys and De Wals, 1981). Clinically, patients suffer tubular dysfunction resulting in aminoaciduria, proteinuria, and glucosuria. Although the half-life of cadmium in kidneys of humans is uncertain, it may be as long as 30 years. Under such circumstances it has been conjectured that critical concentrations [kidney = 200 micrograms (μg) per gram (g) by age 50] could be used to establish maximum levels of daily exposure (Kjellström et al., 1977). What makes cadmium of dietary concern is that ordinary background dietary exposures were estimated to yield kidney concentrations of about one-quarter the hypothesized critical level. The segment of the population at greatest risk would appear to be older adults (ATSDR, 1989b).

Studies in maternal-fetal tissues have provided evidence for accumulation and transplacental transfer of metals. In one study, placental cadmium levels were one to two times those in maternal or cord blood. It was observed also that erythrocyte cadmium levels were roughly three to five times plasma cadmium levels, and that maternal erythrocyte cadmium levels were somewhat higher (27%) than those of the fetus.

Lead

Of all the heavy metals, lead probably has the longest history of environmental contamination and toxicity to humans (Green et al., 1978). For this reason, lead poisoning, or plumbism, has been intensely studied, and a large body of information is available for examination. Sources of lead found in the environment are multiple, and the metal is truly ubiquitous, being commonly found in food, water, and air. Evidence exists that lead in the environment has increased during the past 200 years, and it is not surprising that it can be found as a contaminant of aquatic animals (Shukla and Leland, 1973). Environmental lead is a product of storage battery, ammunition, solder, pigment, pipe, brass, and red lead manufacture. Tetraethyllead is a component of gasoline antiknock additives, although in recent years this use has been drastically reduced. There are at least five pools of lead in the body, two of which reside in the skeleton (90%) in cortical and trabecular bone. Lead in cortical bone is similar in half-life to cadmium (approximately 20 years). Other body compartments for lead include the kidney, lung, and central nervous system (Goyer, 1986). It is not surprising therefore that major lesions and clinical signs in humans suffering frank plumbism are referable to the blood (anemia), brain (convulsions, paralysis), and kidney (proteinuria).

The condition in humans is best known because of its chronic toxicity to young children who ingest lead-base paint chips or lead in soil, house dust from paint, industrial dust, and automotive emissions. Oral ingestion of inorganic lead is without doubt the primary port of entry into humans. Of the lead ingested, only 5-15% is absorbed in adults but considerably more in children (Goyer, 1986).

Recent studies suggest that very low levels ingested by pregnant women may result in learning and behavioral disabilities in neonates and preschool children (Waternaux et al., 1989). Excretion is primarily via the bile and the gastrointestinal tract. Organic lead compounds such as tetraethyllead may be absorbed in large quantities through the skin, but as toxicants these forms of lead are primarily a problem in the petroleum industry. All forms of lead toxicity are less frequent in adults; any occurrence is usually acute and occupationally related (Green et al., 1978).

Lead's toxicological mode of action depends on its molecular configuration, inorganic lead being less toxic than and producing clinical signs different from tetraethyllead. Inorganic lead is an inhibitor of aminolevulinic acid dehydratase (ALAD) and heme synthetase, which leads to anemia (Hammond, 1978).

The metal causes necrosis of neurons, myelin sheath degeneration, and especially, brain vascular damage with increased cerebrospinal fluid (CSF) pressure. These lead to encephalopathy and eventual mental retardation in children. Lead crosses the placental barrier, and there is a good correlation between maternal and fetal blood lead values (Van Gelder, 1978). Therefore, at primary risk from contaminated seafoods are the fetus and neonates.

Mercury

Mercury exists in elemental form, as monovalent (mercurous) or divalent (mercuric) salts, and methylated. The methylated form is the most toxic to humans (Harada, 1978). Methylmercury is formed in the environment from the divalent salts by anaerobic bacteria. It is quite easily absorbed after ingestion and has a variable half-life of 60-120 days in man but is reported to have a half-life of up to 2 years in fish, where it is the predominant form (Al-Shahristani and Shihab, 1974; Stopford and Goldwater, 1975).

The metal is known to produce c-mitosis and chromosomal alterations resulting in cellular damage, with the kidney and brain as target organs. Neuronal damage and axonal demyelination result in the clinical signs and symptoms of paresthesia, incoordination, tremor, and epileptic seizures. The metal also binds strongly to sulfhydral groups (mercaptans), thereby inactivating certain enzymes (Hammond, 1978).

In its methylated form, mercury quite easily passes the placental barrier, placing the fetus at particular risk (Amin-Zaki et al., 1979). The relationship of clinical signs in humans to blood, hair, and urine mercury levels has been reviewed (Tollefson and Cordle, 1986). Children of symptom-free pregnant and nursing mothers with relatively low blood and hair levels may suffer from mental retardation.

Selenium

Selenium is an enigmatic metal because it functions both as an essential nutrient and, at slightly higher levels, as a poison. It is present in various enzymes, has been reported to possess anticarcinogenic effects in animals, is an antioxidant, and yet is a well-documented toxicant of domestic animals as well as a mutagen (Griffin, 1979; Schamberger, 1985; Schnell and Angle, 1983). As an animal toxicant it is a regional problem of the Southwest and Far West. Seleniferous (alkaline, oxidizing) soils give rise to high levels in selenium accumulator plants that are grazed by cattle, sheep, horses, and swine. Poisoned animals develop conditions known as "alkali disease" [subacute, <50 parts per million (ppm)] and "blind staggers" (acute, >100 ppm). Signs include anorexia, tooth and hair loss, watery diarrhea, lassitude, progressive paralysis, and eventual death (Harr and Muth, 1972).

Selenium levels in water from seleniferous areas are often quite high so it is not surprising that selenium has been found as a contaminant of fresh and marine aquatic animals. Its source however is not solely natural. Anthropogenic contamination occurs and is the product of fossil fuel combustion (fly ash) and of paint, alloy, photoelectric battery, and rectifier manufacture (Fishbein, 1983; Sorensen et al., 1984).

Selenium exists in a number of chemical forms, elemental selenium (Se^{o}), selenide (Se^{2+}), selenite (Se^{4+}), and selenate (Se^{6+}). These forms may bond with other metals or organic substances such as amino acids (Ewan, 1978).

The selenates are most soluble and easily enter biological systems. In one study, approximately 15-30% of the selenium found in fish muscle was the selenate form (Cappon and Smith, 1981). The selenites and elemental selenium are relatively insoluble.

This is not to say that selenite when ingested will not act as a toxicant, merely that its innate insolubility may affect its absorption and distribution within the body (Goyer, 1986). The biochemistry of selenium is poorly understood but has been reviewed recently (Reddy and Massaro, 1983).

The mode of action of selenium as a toxicant at the cellular and biochemical levels is uncertain. The metal appears to damage endothelium selectively, resulting in edema and hemorrhage in both humans and animals. It is also responsible for toxic hepatitis with eventual fibrosis (not constituting cirrhosis) in chronic exposures.

Selenosis in animals is reported to produce infertility and congenital malformations (Harr, 1978). Selenosis in man appears to be a relatively rare occurrence, most often due to acute occupational exposure or chronic exposure to contaminated water or food sources. There appears to be very little information regarding the effect on man of chronically high levels of selenium in the diet and its potential risk (Wilber, 1983). Recently, however, levels have been reached in fish that have prompted health alerts in California (Fan et al., 1988).

Organic Compounds

In this section, some of the potential contaminants of seafood that have come to the committee's attention, and about which there are at least some minimal data, are surveyed. These include the chlorinated hydrocarbon pesticides that came into widespread use in the United States and elsewhere immediately after World War II (Hansen et al., 1985). Among the chlorinated hydrocarbon pesticides detected in seafood were benzene hexachloride (BHC) or hexachlorobenzene (HCB), chlordane, dieldrin, dichlorodiphenyltrichloroethane (DDT), endrin, heptachlor, lindane, nonachlor, octachlor, and pentachlorophenol. In addition, industrial chemicals and by-products such as PCBs and dioxins are routinely detected in seafood. Less frequently detected pesticides included chlorpyrifos, dacthal (DCPA), diazinon, ethylene dibromide (EDB), malathion, mirex, omethoate, pentachloroaniline, tecnazene, and trifluralin (FDA, 1988; Gunderson, 1988). In quite specific circumstances, such as in farm ponds in heavily agricultural areas, other chemicals — even those that are not known to bioconcentrate, such as atrazine — can be found in fish (Kansas DHE, 1988). Some pesticides detected are specific to various regions. The carboxylic acid herbicide 2,4-(dichlorophenoxy)acetic acid (2,4-D) has been found in oysters from the northern Chesapeake Bay and Alaskan bivalves (NOAA, 1988). Fish from the Arroyo Colorado and adjacent lower Laguna Madre in Texas contained measurable concentrations of pesticides such as ethion, carbophenothion, ethyl parathion, and methyl parathion (NOAA, 1988). The organic compounds classified here have been reviewed by Murphy (1980).

Specific Organics

Polychlorinated Biphenyls (PCBs)

Polychlorinated biphenyls include more than 200 different compounds ("congeners") that were used in various formulations as liquid insulators in electrical equipment, as encapsulating agents, in carbonless carbon paper, and in hydraulic fluids. The use of PCBs in "open" applications such as carbonless carbon paper was phased out in the early 1970s, and any new use for the remaining applications was stopped in the late 1970s with the passage of the Toxic Substances Control Act. The U.S. usage of approximately 500,000 tons of PCBs in 1930-1970 accounted for about half of the total world production. However, the unusually slow rate of environmental degradation of the more highly chlorinated PCBs in the environment and in higher organisms, and slow continued discharge of PCBs from old equipment and dump sites, have led to a relatively slow rate of decline of PCB concentrations in fish from large freshwater bodies (e.g., the Great Lakes).

The PCBs are a paradigmatic case for the phenomenon of bioconcentration. The more highly chlorinated congeners in particular tend to be both highly lipophilic and very slowly degraded by most organisms. Thus, PCBs that are passed "up" the food chain tend to become much more concentrated as predators are consumed by successively larger predators. In contrast, terrestrial animals that are used for human food are generally vegetarians (first-level consumers of the primary producing organisms).

The principal potential health concerns from PCB exposure include carcinogenesis (on the basis of extensive animal evidence and some suggestive findings in human epidemiological studies), changes in human birth weights, and some loss of neurological performance in the offspring of mothers with relatively high dietary exposures or body burdens (Bertazzi et al., 1987; Brown, 1987; Cordle et al., 1982; Fein et al., 1984; Gladen et al., 1988; J.L. Jacobson et al., 1989, 1990; S.W. Jacobson et al., 1985; Rogan et al., 1986; Sunahara et al., 1987; Taylor et al., 1989).

All carcinogens–in particular, the PCBs and dioxins–are not thought to act primarily by causing DNA mutations (Safe, 1989). Suffice it to say here that lack of knowledge of the precise mechanisms by which PCBs cause cancers makes quantitative assessment of their cancer risk more uncertain than is usual for other chemicals. The PCB mixtures that are delivered to humans via seafood are likely to be systematically different from the original mixtures that were used in animal testing because of more rapid degradation of some (particularly less chlorinated) congeners in the environment and in aquatic organisms. The selection for relatively persistent congeners in aquatic organisms might tend to increase human risk relative to that expected from a naive extrapolation; other factors might have the opposite effect. In any event, the numbers of cases that could be expected seem large enough to warrant exploration of further options for risk reduction.

Dioxins

2,3,7,8-Tetrachlorodibenzo-*p*-dioxin (hereafter known as TCDD) is a contaminant of products made from trichlorophenol, including some chlorophenoxy herbicides. In humans, its effect has been linked to a severe dermatitis; fetal toxicity and numerous other effects have been observed in experimental animals at very low doses.

In standard animal test systems, it is one of the most potent carcinogens known. Using its standard procedures for cancer potency estimation and certain consumption estimates, EPA estimated a lifetime cancer risk of approximately 1 in 100,000 from eating fish contaminated at the nominal detection level of 1 part per trillion (EPA, 1987). Using considerably different methodology for assessing the risk, FDA has advised that, for consumption patterns and species typical to the Great Lakes area, fish consumption should be limited if concentrations in the edible portion exceed 25 parts per trillion and should be banned if concentrations exceed 50 parts per trillion (Kociba et al., 1978).[4] These profound differences in risk assessment indicate the tremendous uncertainty about the true potency of TCDD to cause human cancer.

Polycyclic Aromatic Hydrocarbons (PAHs)

Polycyclic aromatic hydrocarbons are common environmental contaminants found in petroleum, soot, or tar from incomplete combustion, lubricants, and domestic sewage. Many are well-established carcinogens and are highly toxic. Their pervasiveness in the environment assures widespread contamination of aquatic organisms. Because they are poorly metabolized by bivalves and are easily accessible to these animals, they may pose important potential hazards to humans.

Chlorinated Hydrocarbon Pesticides

Like PCBs, the broad group of relatively lipid-soluble, persistent chlorinated hydrocarbons was largely phased out of production in the United States during the 1970s because of concerns for carcinogenicity and ecological effects. Fortunately, few members of the group have proved to be as persistent as PCBs in the environment. Like PCBs, however, the precise mechanisms of action of many chlorinated hydrocarbons in causing cancer appear not to be by direct or indirect reactions with DNA; accordingly, quantitative assessments of risk for this group are more uncertain than usual.

DDT and metabolites: Both DDT and its metabolites [primarily dichlorodiphenyldichloroethane (DDE)] are persistent (slowly eliminated from organisms) lipophilic substances of uncertain health significance in humans, and are among the most widespread and frequently sampled of the chlorinated hydrocarbons.

They are also persistent in ecosystems and bioaccumulate at higher levels of the food chain, resulting in toxicity to birds and aquatic organisms. The use of DDT was essentially banned in the United States in December 1972. Subacute effects of these chemicals at high doses in humans include central nervous system signs and, in rodents, liver toxicity and estrogenic effects. In addition, DDE has been observed to cause liver tumors in rodents.

Dieldrin: Dieldrin (an epoxide of aldrin) is a cyclodiene insecticide that, like DDT, affects the central nervous system, but is more toxic and has caused human fatalities. It too is lipophilic and may be released from fat stores long after exposure, to cause toxicity. It has led to increased liver tumors when fed at relatively low levels to rodents.

Chlordane: Chlordane is similar in molecular structure and mode of action to dieldrin, but is less toxic.

Heptachlor and heptachlor epoxide compounds: Heptachlor and heptachlor epoxide are also chlorinated cyclodienes, and the epoxide is known to be stored in human fat. They have toxicity similar to dieldrin.

Endosulfan: Endosulfan is a cyclodiene pesticide and a problem contaminant in estuaries near agricultural drainage areas due to its widespread use (NOAA, 1989).

Endrin: Endrin is similar in its toxic effects to other cyclodiene pesticides and is more acutely toxic than DDT.

Chlorinated benzenes and phenols: Lindane (^{3}-isomer of 1,2,3,4,5,6-hexachlorohexane), also known as benzene hexachloride (BHC), a mixture of ±-, ^{2}-, and ^{3} -isomers of 1,2,3,4,5,6-hexachlorocyclohexane depending on the manufacturer, is a neurotoxin but has also been found to cause aplastic anemia in humans.

Hexachlorobenzene has never been manufactured in the United States, but it is a ubiquitous fungicide and contaminant often found in other pesticides such as pentachloronitrobenzene (PCNB), which is used in the United States.

Pentachlorophenol: Also known as PCP and penta, pentachlorophenol is a wood preservative, slimicide, and metabolite of the fungicide hexachlorobenzene. Like other polychlorinated phenols, it is contaminated with carcinogenic dioxins (NOAA, 1988).

Mirex: Mirex is a pesticide used to control the fire ant in the southeastern United States. Like other organochlorine pesticides, it is lipophilic and has been reported to be a carcinogen on the basis of rodent studies. It may be a precursor of chlordecone (kepone), is persistent in the environment, and bioconcentrates in the food chain.

Kepone: Kepone has produced appreciable toxicity in exposed workers. It can cause neurological lesions, liver damage, and reproductive failure and is similar in its bioconcentration properties to mirex.

Toxaphene: Toxaphene is a very common domestically used insecticide of complex and often uncertain molecular structure. It is made by chlorinating a mixture of terpenes. It therefore may vary in toxicity from batch to batch depending on the proportion of its isomers. Fortunately, it is of relatively low persistence in the body. Carcinogenic activity is suspected.

Carboxylic herbicides: The herbicides DCPA, 2,4-D, and 2,4,5-trichlorophenoxyacetic acid (2,4,5-T) are chlorophenoxy compounds commonly used in agriculture, and by government agencies and utility companies to control woody plants in rights of way and along roadsides. Although 2,4,5-T has been found to have teratogenic activity, there has long been discussion about whether this is attributable to small amounts of dioxin contaminants. Aside from teratogenesis, these agents can affect animals by impairing neurotransmission, resulting in muscle weakness, ventricular fibrillation, and neuritis.

Atrazine: Atrazine is a herbicide commercially known as AAtrex. It is of low oral acute toxicity. However, its chronic effects are unknown. Structurally similar compounds have been shown to produce thyroid adenocarcinomas in rodents.

Contamination Problems in Aquaculture

Fish culture uses a variety of chemicals that represent potential threats to the health of the cultured animal, indigenous biota, and even the human consumer (Meyer and Schnick, 1989). A number of chemicals of potential toxicity to humans that are not registered for use in the United States are employed in other nations (Fox, 1990). These include furazolidone, nitrofurazone, carofur, chloramphenicol, and silvex–all of which are known or suspected carcinogens.

Chemicals employed in aquaculture include (1) drugs used to treat disease (chemotherapy), (2) those introduced through construction materials, (3) chemicals to treat parasites (formaldehyde), (4) hormones used to alter reproductive viability, sex, and growth rates, and (5) water quality treatments (copper compounds). Of these groups, those of greatest potential concern are the chemotherapeutic drugs. Chemicals used in construction and hormones are not considered in this section because they are relatively nontoxic or have been considered under other headings (organics, pesticides).

Disease is a limiting factor in the culture of aquatic animals. In recent years, culturists have dealt with this problem by developing rapid, immunological diagnostic tests, followed by treatment with drugs commonly used in veterinary medicine. Such chemotherapeutic drugs include, but are not limited to, the potentiated sulfonamides, antibiotics, and nitrofurans. Their widespread use, both nationally and especially internationally, may pose a threat to human health if residues persist in the edible portions of fish and shellfish. Both residue tolerance levels and withdrawal times (the period prior to slaughter during which no drug may be administered) have been determined and set by FDA for major drugs used by veterinarians in the treatment of terrestrial food producing animals.

Hence, the problem of persistent residues has long been recognized by the medical profession, but only recently has it been investigated in cultured aquatic animals. The Center for Veterinary Medicine of FDA is aware of the seriousness of this problem and is actively attempting to deal with it, as evidenced by its recent requests for proposals in the April 9, 1990 issue of *Business Commerce Weekly* (Guarino et al., 1988; Mitchell, 1989).

The proposals will study the pharmacokinetics and methods of monitoring some six drugs, including malachite green and chloramphenicol, presently used in fish and crustacean culture. In addition to this organization, active research is being conducted by the Fisheries Research Branch of the FDA and the Fisheries Research Center, La Crosse, Wisconsin of the U.S. Fish and Wildlife Service (Schnick, 1988). The latter organization has published a booklet entitled "A Guide to Approved Chemicals in Fish Production and Fishery Resource Management" (Schnick et al., 1989). Although there is much interest in the potential toxicity of chemotherapeutic drug residues in seafood, to date only a potentiated sulfonamide (Romet-30), oxytetracycline (Terramycin), and Formalin are approved for use in food producing aquatic animals.

The nature, use, and residues of the drugs applicable to aquaculture that are used to control disease in veterinary medicine have been reviewed (Bevill and Huber, 1977; Booth, 1977; Huber, 1977a-c; Michel, 1986).

Chemotherapeutic drugs used in aquaculture are sulfonamides, antibiotics, and drugs used in the chemotherapy of protozoal, mycotic, and helminthic infections.

In addition to these examples of deliberately used chemicals, fish raised in aquaculture are also susceptible to contamination via pesticides present in feed, agricultural runoff water, and sediments. The magnitude of human exposure to these sources has not yet been assessed and should be examined periodically in light of the growth and change in this sector of the seafood industry.

Sulfonamides

The sulfonamides are readily synthesized derivatives of sulfanilic acid. They are bacteriostatic drugs, and their efficacy may be increased by the addition of ormetoprim (Plakas et al., 1990). Commonly, sulfadimethoxine is potentiated with ormetoprim (Romet-30) and fed to cultured fish at the rate of 50 mg/kg body weight for 5 days. Similarly, sulfamethoxazole is potentiated with trimethoprim (Tribrissen) and fed at the rate of 1-4 mg/kg body weight for 10-14 days. The latter drug is not approved for use in aquaculture. The FDA tolerance level for both these drugs in cattle and chickens is 0.1 ppm (Booth, 1977). The sulfonamides have been implicated in renal damage, urinary obstructions, and hematopoietic disorders. For that reason, FDA has set withdrawal times and residue tolerance levels to protect public health. The withdrawal period for catfish is 3 days and for trout 6 weeks (Schnick et al., 1989).

Antibiotics

The term antibiotic includes a large number of drugs inimical to the growth of microorganisms. Of greatest interest in aquaculture are oxytetracycline, ampicillin, chloramphenicol, and erythromycin.

Oxytetracycline (Terramycin) is a product of the mold *Streptomyces rimosus*. It is a broad-spectrum antibiotic used to treat a number of bacterial diseases in fish and crustaceans. It is the only antibiotic approved by the FDA for use in aquaculture. Contamination of the food supply with the drug is a public health concern because chronic exposures at low levels may lead to a higher incidence of antibiotic-resistant bacterial strains, poor growth of teeth, and the possibility of photosensitivity. The FDA tolerance level for oxytetracycline in meat is 0.25 ppm, whereas in fish it is 0.1 ppm with a withdrawal time of 21 days. Elimination time from muscle appears longer for salmonids than for catfish and is dependent on the water temperature (Plakas et al., 1988).

Ampicillin is a newer, semisynthetic penicillin that has activity against both gram-positive and gram-negative organisms. Although not approved for use in the United States and seldom used domestically, it is commonly used in Japan to control pasteurellosis in yellowtail culture (Hawke et al., 1987). Ampicillin, like its congeners, may result in severe hypersensitivity reactions in some people (Huber, 1977a). Because an initial exposure to some form of penicillin is necessary to produce an eventual drug allergy, ingestion of ampicillin as a residue in seafood is of public health interest. The FDA has set a tolerance level of 0.01 ppm and a preslaughter withdrawal time in cattle of 6 days. No information is available regarding residues or withdrawal time in fish or crustaceans.

Chloramphenicol is a product of the mold *Streptomyces venezuelae* and is a potent antibiotic that is effective against most bacteria, as well as rickettsia and the psittacosis-lymphogranuloma group of organisms. Development of resistant bacteria may follow its widespread and uncontrolled use. Chloramphenicol administered for long periods may cause blood dyscrasias such as aplastic anemia and has recently been incriminated as a carcinogen. It is not approved for use in the United States in food producing animals. The drug is used, however, with impunity to control bacterial diseases of shrimp, especially in Ecuador and in European fish culture (Manci, 1990; Meyer and Schnick, 1989). Because imported seafood is not checked for antibiotics, no information is available regarding its residue levels.

Erythromycin is a product of the mold *Streptomyces erythreus* and is primarily effective against gram-positive organisms. The drug is not approved for use in domestic fish culture but is used in Europe. The tolerance level in U.S. swine is 0.01 ppm, and the withdrawal time is 7 days. The drug is not considered an important cause of hypersensitivity in man and is relatively nontoxic.

Nitrofurans

Nitrofurans are synthetic compounds that are active against most gram-negative bacteria, some fungi, and some protozoal organisms. Because of their toxicity, their clinical use is limited and FDA has attempted unsuccessfully for over 15 years to ban their use in food producing animals and to have them withdrawn from the market (FDA, 1982). Side effects include bleeding, gastrointestinal upsets, and allergic reactions. Tolerance levels for the drug have been set at zero. In aquaculture, their use has been suggested for the treatment of *Ichthyophthirius multifillis*.

CONCLUSIONS

1. Certain drugs with potential toxicity to humans are used to control disease in cultured food producing aquatic animals.
2. There is a paucity of information regarding the withdrawal times, residue levels, and pharmacokinetics in the cultured aquatic animals receiving these drugs.
3. There is some reason for concern that large amounts of imported cultured seafood, which is routinely treated with drugs, are consumed by the American public, although the magnitude of ultimate human exposure from this source is as yet uncertain.
4. Cultured seafood imported into the United States is not presently inspected for drug residues.

Contaminants as a Result of Processing

Nitrosamines

Nitrosamines are formed in smoked fish products and in the human stomach as the result of the simultaneous presence of secondary amines and nitrite. Since the late 1960s, the use of nitrite in smoked fish has been authorized to avoid the repetition of botulism incidents in the early 1960s that resulted from the mishandling of vacuum packed smoked fish. [Earlier joint industry/FDA recommendations on the time and temperature cooking of smoked fish products were reportedly not complied with by Great Lakes producers, and the food additive petition was later granted after it was shown that nitrosamine formation in the fish itself did not exceed a specific level of sensitivity specified by FDA (Hattis, 1972).] The exact extent of the extra exposure to specific nitrosamines attributable to the use of nitrite as a food additive in smoked fish has apparently not been reappraised recently by using more sensitive analytical techniques and current risk assessment procedures.

Products of Chlorination, Bromination, and Iodination

Chlorine and some other active halogen compounds are widely used as disinfectants in seafood processing, as well as in the treatment of drinking water and sewage effluents (Fukayama et al., 1986). This process is known to generate some levels of halogenated amines, aromatics, and methanes (e.g., chloroform), and at least one test in chiller water used in chicken processing detected mutagenic activity when the chlorination level was raised to 250 ppm (Masri, 1986). The extent of contamination of seafood with products resulting from the use of chlorine and other halogen compounds does not appear to have been assessed, and there are no relevant appraisals of the associated risks in the available literature. Some relevant information may be contained in food additive petitions for disinfectants that have been submitted to FDA, but the committee has not obtained these documents.

Residues of Ozonation

Ozone treatment is frequently used in foreign settings as a method of depuration of some shellfish (Fauvel et al., 1982). This technology has also been introduced into the United States for use in icing fish (Rice et al., 1982), washing seafoods, and cleansing saltwater for use in molluscan shellfish depuration. In recent years, FDA has raised questions about the safety of this process because of the likelihood of residual reaction by-products of the resulting oxidation. This concern was recently debated during the First International Molluscan Shellfish Depuration Conference, November 5-8, 1989 in Orlando, Florida

(W.S. Otwell, University of Florida, Gainesville, personal communication, 1989) and later in letters from the FDA Compliance Branch (J.A. Baca, FDA, personal communication, June 22, 1990) explaining its position on the use of ozone in food manufacturing facilities. The primary concern is any ozone contact with seafood whereby the ozone could become a component of the food or affect the character of the food. The prevailing differences of opinion await more technical resolution in studies to monitor the consequences from direct and indirect (treating depuration water) applications. Cognizant of the current data, the FDA regulatory interpretations to restrict use will most likely prevail.

Sulfites

Sulfites have traditionally been used to prevent melanosis in crustaceans (Camber et al., 1956). Sanctioned procedures include a 1-minute dip in concentrations of up to 1.25% sodium bisulfite or metabisulfite. This has been shown to impart a residual sulfite level of less than 100 ppm on the edible portion of penaeid shrimp.[6] Product treated in this way must be labeled to designate prior use and residuals in excess of 10 ppm. General concern for allergic-like reactions, particularly in some asthmatic people, has given rise to concern about the continued use and appropriate labeling of sulfites (Lecos, 1985). No equally effective alternative processes have yet been developed for crustaceans.

2

Health Risk Assessment

The difficult issue of determining human intakes for a broad (though far from comprehensive) range of chemical contaminants in seafood is subsequently addressed. Estimates are made of national average daily intakes of various inorganic and organic contaminants via commercially marketed seafood, and for several organic carcinogens, upper-confidence-limit estimates of possible cancer risk are made. However, aside from the methylmercury example,[1] in the absence of better information on the population distribution of the dosage of contaminants to the U.S. population, it is impossible to make even tentative quantitative estimates of potentially significant noncancer risks. Of additional serious concern are the appreciable quantities of seafood consumed following noncommercial sport and subsistence tribal fishing.

Finally, an overview of opportunities for research on different categories of potential health impacts is presented, and conclusions are drawn from both this and the previous chapter. The principal conclusions are the following:

From both natural and human sources, a small proportion of seafood is contaminated with appreciable concentrations of potentially hazardous organic and inorganic chemicals. Some of the risks that may be significant include reproductive effects from PCBs and methylmercury, and carcinogenesis from selected PCB congeners, dioxins, and some chlorinated hydrocarbon pesticides.

Consumption of some types of contaminated seafood poses enough risk that efforts toward evaluation, education, and control of that risk must be improved. Present quantitative risk assessment procedures used by government agencies can and should be improved and extended to noncancer effects.

Current monitoring and surveillance programs provide an inadequate representation of the presence of contaminants in edible portions of domestic and imported seafood, resulting in serious difficulties in assessing both risks and specific opportunities for control.

Because of the unevenness of contamination among species and geographic areas, it is feasible to narrowly target control efforts and still achieve meaningful reductions in exposures.

The data base for evaluating the safety of certain chemicals that find their way into seafood via aquaculture and processing is too weak to support a conclusion that these products are being effectively controlled.

The principal recommendations of the committee are as follows:

Existing regulations to minimize chemical and biological contamination of the aquatic environment should be strengthened and enforced.

Existing FDA and state regulations should be strengthened and enforced to reduce the human consumption of aquatic organisms with relatively high contaminant levels (e.g., certain species from the Great Lakes with high levels of PCBs, swordfish and other species with high methylmercury levels).

Federal agencies should actively support further research to determine the actual risks from the consumption of contaminants associated with seafood and to develop specific approaches for decreasing these risks.

Increased environmental monitoring should be initiated at the state level, as part of an overall federal exposure management system.

States should continue to be responsible for site closures, and for issuing health and contamination advisories tailored to the specific consumption habits, reproductive or other special risks, and information sources of specific groups of consumers.

There should be an expanded program of public education on specific chemical contaminant hazards via governmental agencies and the health professions.

INTRODUCTION

Part of the committee's charge was to review and summarize "the current status of regulations, guidelines, and advisory statements issued by Federal and State public health authorities on environmental contaminants in seafood." Its review was to specifically address contaminants defined by Food and Drug Administration (FDA) regulations as "avoidable or unavoidable." Then, based on this, the committee was asked to "assess how well the current regulatory framework protects the public health."

The committee was also charged with the task of reviewing and summarizing, specifically, the health risk assessment procedures used by FDA, the Environmental Protection Agency (EPA), and other regulatory authorities for priority environmental pollutants, including toxic metals and synthetic organic chemicals. In addition, the committee was asked to "recommend future research directions, as appropriate."

To set the stage for an examination of how current risk assessment procedures can be improved, the basic concepts underlying the mechanisms of action of toxic substances are articulated in the following section, along with quantitative ideas about dose-time-response relationships. Then an extensive critique of agency risk assessments for PCBs and methylmercury is provided. Finally, the committee addresses issues of exposure assessment and risks from other substances, and opportunities for further research on potential chemical health hazards.

BROAD CATEGORIZATION OF MECHANISMS OF DIFFERENT ADVERSE EFFECTS AND IMPLICATIONS FOR DOSE-RESPONSE RELATIONSHIPS

The concern from which much of our regulatory history has resulted is the potential carcinogenic effect of some contaminants. To a certain extent, this concern is based on the mutagenic mechanisms of cancer. For carcinogens that act by primary genetic mechanisms, there are good theoretical reasons to believe that at the limit of low dosage, the risk will be a linear function of exposure (Ehrenberg et al., 1983; Hattis, 1990a). However, as more is learned about the mechanisms of some other types of toxic effects–particularly reproductive effects and chronic degenerative neurological conditions (NRC, 1989, 1990)–concern about potential low-dose effects of other types has tended to increase. It is therefore important to clarify what the expectations should be for dose-response relationships from first principles, given the full range of causal processes that can lead to impairment of health.

Are the events occurring ordinarily fully reversible (or very nearly so), given a prolonged period with no further exposure to the hazard?

Types of Health Hazards Requiring Fundamentally Different Risk Assessment Approaches

1. "Traditional" toxicity resulting from overwhelming body compensatory processes: below some threshold, in individuals who are not already beyond the limits of normal function without exposure, response is reversible.

- Traditional acute toxicity–Toxic action is completely reversible or proceeds to long-term damage within about three days of exposure (paralytic shellfish poisoning, puffer fish poisoning; probably many teratogenic effects).
- Traditional chronic toxicity—Toxic process typically proceeds to permanent damage over a period of several days to several months, due to either (1) reversible accumulation of a toxic agent (e.g., methylmercury, lead) or (2) accumulation of a slowly reversible toxic response (e.g., cholinesterase inhibition).

2. Effects resulting from insidious processes that are irreversible or poorly reversible at low doses or early stages of causation.

- Molecular biological (stochastic process) effects—Effects occur as a result of one or a small number of irreversible changes in information coded in DNA: mutagenesis, most carcinogenesis, and some teratogenesis.

- Chronic cumulative effects–Effects occur as a result of a chronic accumulation of many small-scale damage events: emphysema, noise-induced hearing loss, atherosclerosis, and probably hypertension; possibly depletion of mature oocytes.

If the answer to the first question is yes, then it will generally be appropriate to treat the condition within the framework of traditional toxicology.[3] Some examples of such reversible changes are the following:

Buildup of a contaminant in blood or other tissues. It is rare for there to be a zero rate of excretion of any material. Given time and no further exposure, toxicant buildup should be reversible, although it can be quite prolonged. [Current estimates are that only about 9% of more persistent polychlorinated biphenyl (PCB) isomers may be metabolized or excreted per year in humans (Yakushiji et al., 1984).]

Most enzyme inhibition (generally, even irreversible inactivation of enzyme molecules can be "reversed" through the synthesis of replacement molecules)

Induction of short-term biological responses that act to maintain homeostasis (e.g., sweating in response to heat, tearing in response to eye irritation)

If the answer to the above question is no and events are likely to be occurring at subclinical exposure levels or preclinical stages that are not ordinarily reversible, the modeling of biological risks will have to be based on concepts that are fundamentally different from the homeostatic system/threshold paradigm. Examples of such irreversible or poorly reversible events include:

- changes in genetic information or the heritable pattern of gene expression after these are effectively "fixed" into a cell's genome expression by replication;
- death of nonreplicating types of cells (adult neurons);
- destruction of nonregenerating structures (alveolar septa); and
- generation and buildup of incompletely repaired lesions (atherosclerotic plaques).

Appropriate modeling for conditions that are the result of irreversible or poorly reversible processes must be based fundamentally on the likely dose-response characteristics of the events that cause the basic irreversible changes. Once the primacy of such changes is established for a particular event, the analyst should then ask whether clinical manifestations are likely to be the direct result of only a few, or very many, individual irreversible damage events. If only a few events are believed to contribute directly to a particular clinical manifestation (e.g., a small number of heritable changes within a singe cell line leading to cancer), the effect can be considered a "molecular biological" disease. The risk assessment models used must follow from an understanding of the stochastic nature of the basic process. On the other hand, if thousands, millions, or billions of individual irreversible events directly contribute to a particular condition (e.g., very large numbers of individual neurons must die to cause the clinical manifestations of Alzheimer's or Parkinson's disease), the biological harm should be dealt with under the novel category of chronic cumulative conditions (see below).

Traditional Acute Toxicity

Three kinds of insights for acute toxicity risk assessment follow naturally from the homeostatic system paradigm of physiology and traditional toxicology:

1. There will be a series of toxic effects as different compensatory processes are overwhelmed and as impairment broadens from more-to less-sensitive cells and functions.
2. For each effect in each individual who is not already beyond the limits of normal functioning in the absence of exposure, there will be some subthreshold level of exposure that will be insufficient to produce the effect.
3. Individuals will differ in their thresholds.

A caveat to the general expectation of individual thresholds is that some tasks may so tax the capabilities of a system (perhaps, during fetal life, the struggle to mobilize metabolic resources to grow and differentiate as fast as possible so as to cope with the external world at birth) that any impairment of a key limiting functional parameter required for the task could compromise function to some degree. (This would also apply to reaction time for a driving task, for example.) Of particular relevance to the committee's task in this regard is the suggestion of some studies that dietary PCB exposure may be associated with either changes in birth weight (Fein et al., 1984; Sunahara et al., 1987; Taylor et al., 1989) or indices of neurological function in infants (Jacobson et al., 1985; Rogan et al., 1986).

The first job in assessing acute toxic effects is to define the series of acute responses to the disturbing influence in question. Ideally, the analysis should then attempt to determine (to whatever degree of precision is possible) the nature and magnitude of the dosage and the disturbance of physiological parameter(s) that are necessary to cause each type of acute toxic response, along with the frequency of each response in a diverse human population.[4]

Such mechanism-based analysis is, however, not common in the field. Rather, the current state of the art in those rare cases where acute toxic effects are treated quantitatively is to use probit equations (Finney, 1971) of the general form

Prohibit of response = a + b ln (Cn T),

where *a*, *b*, and *n* are constants, *C* is external concentration, and *T* is exposure time; *n* represents the basic trade-off between intensity and duration of exposure, and *b* defines the breadth of an assumed lognormal distribution of threshold responses. Although some of the available animal data on irritant gases appear to be well summarized by equations of this form (Appelman et al., 1982), this is basically an empirical formula that does not incorporate quantitative representations of the various processes underlying toxicity. It is therefore difficult to decide what adaptations should be made in applying the empirical relationships to diverse subsets of humans.

Even more common, unfortunately, is the simple use of the no-effect level (NOEL)/"safety factor" analysis for arriving at acceptable daily intake (ADI) levels for chemical contaminants. Rather than estimate the numbers of people with specific degrees of particular effects, the general approach is to arrive at an ADI by a rule-of-thumb procedure derived from observed NOELs [or, more recently, no-adverse effect levels (NOAEL), after often contentious discussion over what effects are considered "adverse"], or sometimes low effect levels (LOELs) in animal experiments or human studies. When projections are made from animal data, generally a 100-fold "uncertainty factor" is allowed between the NOEL and the ADI (Dourson and Starra, 1983). The 100-fold factor is often decomposed as 10-fold to account for possible differences in sensitivity between humans and the most sensitive species tested and 10-fold to account for possible interindividual differences in susceptibility among humans.

This approach has a few advantages:

1. It is "quick and dirty"—relatively straightforward to apply and does not require complicated model building or analysis.
2. Through thousands of applications in the past, it is not yet known to have led to catastrophic adverse effects in humans (using Ozonoff's working definition of a "catastrophe" as an effect so large that even an epidemiological study can detect it (D. Ozonoff, Boston University School of Public Health, personal communication, 1990).

On the other hand, for the long term, the simple uncertainty factor approach has a number of disadvantages:

1. No one knows how protective it really is, either in general or in specific cases. What fraction of the diverse human population can be expected to experience adverse effects when exposed at the level calculated to be "acceptable" under the formula? (In general, there may be some finite fraction of individuals who, because of disease or other reasons, are marginal for biological functions affected by the chemical and who may be pushed beyond a functional threshold for an adverse effect by a small finite dose of the chemical.)
2. The procedure incorporates one specific social policy standard for setting "acceptable" levels without making

clear where technical analysis leaves off and policy/ value analysis begins.

3. Effects are generally scored as either present (operationally, statistically significant) or not present (not statistically significant) at a particular dose. There is usually no quantitative analysis of the effects of sample size or the dose-response relationship for the effect in question.
4. There is no defined or obvious way to incorporate newer types of relevant data on human interindividual differences in rates of uptake/absorption for a constant environmental exposure ("exposure variability");

- rates of activating or detoxifying metabolism and excretion, producing differences in the concentration x time of active metabolites per unit of absorbed dose at the site of toxic action ("pharmacokinetic variability"); and
- differential risk of response ("response variability") for a given concentration x time of active metabolites at the site of toxic action.

In particular, the committee suspects that the inability of the uncertainty factor paradigm as usually formulated to incorporate newer types of relevant information into a systematic procedure for updating assessments of health hazards has tended to discourage both the collection and the analysis of potentially important data. One example of this is information on human interindividual variability in parameters that could affect susceptibility.

Such a parameter is generally a continuous variable that has a strong causal influence on performance of an important biological function (although it will not generally be the sole determinant of performance). It should, in turn, be affected by the toxin/exposure under study, and it should be reasonably likely that effects on the final health condition of concern are primarily mediated through effects on this functional intermediate parameter. For example, a key functional intermediate for some reproductive effects of PCBs may well be changes in birth weights (Fein et al., 1984; Jacobson et al., 1985; Sunahara et al., 1987; Taylor et al., 1989). Similarly, blood or tissue concentrations of lead constitute a useful intermediate parameter for lead toxicity.

Elements of a New Analysis for Noncancer Health Effects Mediated by a "Functional Intermediate" Parameter

1. Elucidate the quantitative relationships between internal dose/time of toxin exposure and changes in the functional intermediate parameter.
2. Assess the preexisting "background" distribution of the functional intermediate parameter in the human population.
3. Assess the relationship between the functional intermediate parameter and diminished physiological performance or adverse health effects.
4. Assess the magnitude of parameter changes likely to result from specific exposures in humans (taking into account human interindividual variability in metabolism and other determinants of pharmacokinetics) and consequent changes in the incidence and severity of health effects.
5. Do not attempt, from the biology alone, to determine "acceptable" levels of parameter change or exposure. (Let policymakers decide what changes in the incidence and severity of health effects are "acceptable" in the context of modes of exposure and in light of the feasibility of reducing or avoiding exposure.)

In principle, the use of such intermediate parameters can provide windows on the pathological processes that occur earlier in the development of toxicity, are more sensitive to the action of potential toxicants (compared with attempts to observe actual cases of illness), and are more accessible to direct comparative measurement in both animal models and humans. It is desirable, for these purposes, that the intermediate parameters chosen be as close as possible to the actual causal pathway leading to harm.

However, even a parameter such as birth weight, which may not itself bear a direct causal relation to infant deaths, may be a close enough indicator of the actual causal processes to serve as a useful intermediate predictor.

TABLE: *Expected Infant Mortality Effects of a 1% (33.66-g) Reduction in Birth Weight*

Weight Range (g)	Fraction of Births		Mortality Risk per 1,000 Births in Category	Fraction of Births x Mortality Risks/1,000		
	Original Population	After 1% Weight Reduction		Without Birth Weight Reduction	After 1% Birth Weight Reduction	Net Change
White Infants						
Under 500	0.0006912	0.0007701	1,000	0.6912	0.7701	0.0789
500-999	0.002171	0.002335	673.31	1.4612	1.5720	0.1107
1,000-1,499	0.005249	0.005488	237.85	1.2485	1.3053	0.0568
1,500-1,999	0.009182	0.009575	76.86	0.7057	0.7360	0.0302
2,000-2,499	0.029192	0.032804	26.746	0.7808	0.8774	0.0966
2,500-2,999	0.15164	0.16568	8.3565	1.2672	1.3845	0.1174
3,000-3,499	0.36237	0.37081	4.2566	1.5424	1.5784	0.0359
3,500-3,999	0.31749	0.30337	3.0451	0.9668	0.9238	0.0430
4,000-4,499	0.10100	0.09027	3.0293	0.3060	0.2734	0.0325
4,500+	0.021021	0.01890	4.941	0.1039	0.0934	0.0105
Total	1	1		9.0736	9.5142	0.4406
Black Infants						
Under 500	0.0026095	0.0028661	1,000	2.6095	2.8661	0.2566
500-999	0.006279	0.006666	645.90	4.0558	4.3058	0.2500
1,000-1,499	0.012709	0.013154	167.98	2.1348	2.2096	0.0748
1,500-1,999	0.020673	0.02165	57.72	1.1932	1.2495	0.0563
2,000-2,499	0.067052	0.074351	21.482	1.4404	1.5972	0.1568
2,500-2,999	0.24894	0.26444	9.832	2.4476	2.6000	0.1524
3,000-3,499	0.38248	0.37936	6.636	2.5381	2.5174	0.0207
3,500-3,999	0.20683	0.19100	5.581	1.1543	1.0660	0.0883
4,000-4,499	0.04418	0.04029	5.89	0.2602	0.2373	0.0229
4,500+	0.008253	0.006226	12.33	0.1018	0.0768	0.0250
Total	1	1		17.9358	18.7257	0.7899

Because there will generally be a series of steps in the sequence between toxin uptake and ultimate manifestation of adverse effects, the analyst may often have choices of which parameter(s) to use for assessing human risk. These choices will generally be based on the availability of measurement techniques and theory for observing or estimating the parameter(s) in question.

Traditional Chronic Toxicity

The basic principles that govern the analysis of acute toxic effects are by and large directly applicable to cases of chronic toxicity. Chronic toxic analyses tend to differ from analyses of acute toxic effects primarily in that considerable emphasis must be placed on the "slow step" of the process, which causes the effect to be chronic rather than acute. This slow step is generally either a long-term accumulation of a toxic agent that is poorly excreted under ordinary conditions (e.g., lead, mercury) or an accumulation of some slowly reversed residual effect (e.g., acrylamide).

With lead as an example, the following statements can be made:

Appreciable information is available on the pharmacokinetics of lead absorption, transport, storage, and excretion (Barry, 1975; Bernard, 1977; Campbell et al., 1984; Chamberlain, 1985; Marcus, 1985a-c; Rabinowitz et al., 1976); and much better information could be obtained with the aid of natural experiments such as strikes among lead-exposed workers, which can given information on the rate of decrease of blood lead levels after a reduction in exposure (Hattis, 1981).

Inhibition of heme synthesis enzymes at essentially all dose levels is well characterized (Haeger-Aronsen et al., 1974), and the inhibition of heme synthesis may be important in producing some of the neurotoxic effects of lead (Silbergeld et al., 1982), although the short- and long-term functional significance of different degrees of inhibition in different individuals is far from clear.

Effects on some measures of neurological function and kidney function are susceptible to study in reasonably straightforward ways. Effects on higher-order development of central nervous system functions are more difficult to determine because of an ignorance of basic mechanisms; however, some good studies have become available in recent years (Baker et al., 1983; Bellinger et al., 1987, 1990; HHS, 1988; Needleman et al., 1979, 1990; Waternaux et al., 1989). The impairment of very complex neurological functions by lead raises a significant issue in the application of the traditional toxicological paradigm to risk analysis.

As indicated above, the usual assumption is that there is some functional reserve capacity in "normal" individuals that maintains "adequate" performance despite a "small" degree of perturbation of a biological parameter by a "low" dose of toxic material. However, if the function is already taxed to its limit in certain situations, even in the absence of exposure (perhaps for a first grader learning to read or for a developing fetus mobilizing all its available metabolic energy to grow and differentiate), and if the biological parameter being perturbed is limiting to the performance of that function, then any level of exposure may produce at least some reduction in performance.

Addressing the issues of the population distribution of different functional reserve capacities, and the relationship of functional reserve capacities to specific biochemical parameters, is essential to the future research needs of risk assessment for classical chronic toxic agents. Also in the area of neurotoxicology, Silbergeld (1982) has written of the potential of new radioimmunoassay and functional measurement techniques to help shift the focus of research away from traditional morphological criteria of neurological damage toward more sensitive and sophisticated measures of performance.

Molecular Biological (Stochastic Process) Diseases

In addition to most carcinogenesis, molecular biological diseases include mutagenesis and at least some teratogenesis. The subject of quantitative risk assessment for carcinogenic hazards has been discussed extensively elsewhere (American Industrial Health Council, 1987; Bishop, 1987; Crump et al., 1976, 1977; EPA, 1986a; Hattis, 1982; Hattis and Smith, 1986; Moolgavkar, 1986; Rai and Ryzin, 1981; Whittemore, 1980). However, it is worth briefly recapitulating some basic features of the carcinogenic process and their implications for cancer dose-response relationships.

Science is now much closer than it was a decade ago to understanding the fundamental mechanisms involved in carcinogenic transformation. For some time it has been clear that tumors arise as a result of a series of changes or rearrangements of information coded in DNA within single cells (Cleaver and Bootsma, 1975; Fialkow, 1977; Hattis, 1982; Knudson, 1973, 1977; McCann et al., 1975; Vogel and Motulsky, 1979). These changes are often induced by electrophilic metabolites of the parent compounds to which organisms are exposed (Miller and Miller, 1981). With the identification of "oncogenes," some detailed molecular characterization is being provided of the changes resulting in DNA (Fischinger and DeVita, 1984; Hoel, 1985; Modali and Yang, 1984; Yunis, 1983).

It has also been apparent for some time that further headway cannot be made in elucidating the shapes of carcinogenesis dose-response relationships at low dosages simply by increasing the numbers of animals studied in conventional bioassays. A variety of mathematical models with dramatically different consequences for low-dose risk can always be found that fit the observations about equally well (Maugh, 1978; Whittemore, 1983). Low-dose risk projections are, therefore, inevitably much more determined by the choice of model than by the available data (Guess et al., 1977; Whittemore, 1980), if what is meant by "data" is restricted to observations of the incidence of ultimate adverse effects in small groups of animals.

Because of sample size limitations, animal carcinogenesis bioassays must be done within a limited range of relatively high dose levels. Typically, the difference between the minimum detectable response and a response that effectively saturates the system or causes interference through overt toxicity is only one to two orders of magnitude (often even less). Over this high dose range near levels where the agent produces overt toxic effects, enzyme saturation and other forms of pharmacokinetic nonlinearities are most likely. If in dose-response modeling for risk assessment, the nonlinearities of pharmacokinetic origin are not separated from the nonlinearities that may arise from the multiple mutation mechanism that is central to carcinogenesis, our ordinary curve-fitting procedures will implicitly attribute the pharmacokinetic nonlinearities to the fundamental carcinogenic process (Hoel et al., 1983). The resulting errors are particularly serious if one wishes to produce the best point estimates of carcinogenic risk in addition to upper confidence limits.

Clearly, to make real progress in modeling carcinogenic risks, knowledge of the fundamental processes involved must be used to break open the black box between external exposure levels and ultimate production of tumors. The use of pharmacokinetic models and intermediate parameters ("markers") to characterize the dose-response characteristics of small segments of the causal pathway to carcinogenesis has considerable potential to improve dose-response modeling for the process as a whole (Hattis, 1988). Such markers can include both those that may lie directly along the causal pathway, such as DNA adduct formation, and putative correlates, such as hemoglobin adduct formation, that can be good indicators of the concentration-time product of active intermediates in the systemic circulation.

One key fact must be recognized from the beginning about pharmacokinetic modeling, however. Whatever nonlinearities may be produced at high doses by the saturation of enzymes, the saturation of active transport processes, the depletion of cellular reactants for electrophilic agents, or changes in cell division rates to make up for cell killing due to overt toxicity, all of these nonlinearities must necessarily disappear as one approaches very low dose rates (Hattis, 1990a). The slope of the line relating ultimate DNA lesions in replicating cells to external dose may well be very different at low than at high doses, but it must be linear.

The basic reason for this is that at low doses the rates of the transport and transformation processes that lead to DNA damage and repair depend directly on the number of collisions between molecules of an "input" chemical (or activated intermediate or DNA adduct) and a resident cellular reactant (or hole in a membrane or repair enzyme molecule). At low doses the number of resident cellular reactant molecules does not change appreciably as a function of the concentration of the input. Therefore, the number of relevant collisions and the rates of reactions and side reactions in the causal sequence at low dosage must be direct linear functions of the amounts of input chemical and its activated derivatives. Some finite fraction of the ultimate DNA lesions must escape repair before the next cell replication as long as the cells affected have a nonzero turnover rate, there are a finite number of repair enzyme molecules, and the repair molecules operate at a finite rate.

All carcinogens – in particular the PCBs and dioxins – are not thought to act primarily by causing DNA mutations (Safe, 1989). There is disagreement about how to estimate risks from carcinogens. In the United States, regulatory agencies generally estimate risks in the same way for both genotoxic and non-genotoxic carcinogens. Regulatory agencies adopt this position because they believe that full knowledge of the mechanism of action of non-genotoxic agents is needed before they can be assumed to exhibit thresholds. In several foreign countries, non-genotoxic carcinogens are generally assumed to have thresholds below which there is expected to be no risk. The disagreement is not confined to official government positions; some scientists prefer to treat non-genotoxic carcinogens as having thresholds, some do not.

TABLE: "Indirect" Mechanisms of Carcinogenesis

Indirect processes that enhance the rate of "initiation" (the initial change or rearrangement of information in DNA that places a cell on a pathway to cancer):

1. Changes in basic transport processes (e.g., low-fiber diets may prolong the residence time of feces in the gut, leading to greater exposure of the intestinal epithelium to reactive agents).
2. Changes in metabolic processing (e.g., induction of enzymes such as some mixed function oxidases that convert chemicals to forms that can directly react with DNA).

3. Changes in the effective amount of target tissue available for carcinogenesis (e.g., by simple hyperplasia).
4. Changes in the efficiency of DNA repair [e.g., inhibition of DNA repair by some metals (Zakour et al., 1981) or enhancement of cell replication (leaving less time for repair before a DNA lesion can be fixed into the genome as a permanent mutation)].

Indirect processes that alter the frequency with which "initiated" cells progress through subsequent stages in the carcinogenic process:

1. Induction of subsequent genetic changes: Many promoters appear to be capable of inducing the expression of Epstein-Barr virus antigens (Takada and Zur Hausen, 1984). Additionally, some "promoters" reportedly lead to the generation of active oxygen species that may damage DNA and lead to subsequent somatic mutations along the pathway to carcinogenesis without themselves being converted to compounds that react directly with DNA (Kinsella and Radman, 1978).
2. Changes in the frequency with which initiated cells are effectively removed by terminal differentiation [e.g., effects of early and multiple pregnancies in reducing later breast cancer risk (Kampert et al., 1988; Layde et al., 1989; Moolgavkar et al., 1980)].
3. Release of initiated cells from growth control by neighboring cells
   - Mimicry of the action of a growth regulator or hormone by an introduced substance. [Phorbol esters alter the binding and phosphorylation of epidermal growth factor receptors (Friedman et al., 1984; McCaffrey et al., 1984). The very high-affinity binding of 2,3,7,8-tetrachlorodibenzodioxin (TCDD) to the AH receptor induces an increase in transcription of the genes that code for certain P-450 isozymes, among other effects (Whitlock, 1989). Genetic studies suggest that the AH receptor is necessary but not sufficient for the activity of

halogenated aryl hydrocarbons as skin tumor promoters in hairless mice (Safe, 1989).]

- Inhibition of the action of normal growth suppressing substances. [Phorbol esters inhibit the binding of somatostatin (Zeggari et al., 1985).]
- Inhibition of the passage of growth inhibitors among cells by many chlorinated aromatic compounds (Tsushimoto et al., 1983).
- Killing of neighboring cells responsible for repression of initiated cells.
- Induction of cell replication among initiated cells, interfering with the ability of repressors to pass tight junctions or isolating some daughter cells from tight junctions.

4. Changes in the rates of proliferation or survival (without terminal differentiation) of initiated cells relative to the proliferation and survival of normal cells (theoretical mechanism suggested by Moolgavkar and Knudson, 1981).

Indirect processes that might alter the survival, growth and spread of tumors, or the progression of tumors to increased malignancy:

1. Changes in hormonally mediated processes that might speed up the growth of specific cell types (e.g., estrogens and some breast cancers).
2. Changes in the efficiency of immune surveillance in destroying incipient tumors at early stages. Some observations suggest that tumor promoters may alter the functioning of "natural killer" cells (Kabelitz, 1985). Immunosuppressive effects have also been observed for some promoters in vivo (Pasquinelli et al., 1985).
3. Changes in local tissue conditions that favor colonization of new tissues by metastases (e.g., the establishment of tumor blood supplies).

Generally, U.S. regulatory agencies adopt linear, no-threshold models and do not depart from them unless there is overwhelming scientific evidence to show that they are incorrect in specific cases. This is a prudent posture to take, until evidence of the details of specific nongenetic mechanisms is detailed enough for different low-dose assumptions to be indicated very clearly.

Chronic Cumulative Conditions

Chronic cumulative conditions include neurological conditions caused by the cumulative loss of neurons (e.g., Parkinsonism, Alzheimer's disease), emphysema and other chronic lung diseases, atherosclerosis, and hypertension. This new category is required because the underlying mechanism of these conditions – slow accumulation of many irreversible or poorly reversible damage steps – departs significantly from paradigmatic mechanisms that provide the basis for the three other, more traditional categories. These conditions are likely to be increasingly important as the population shifts to progressively older ages, and as continued progress is made in preventing and treating diseases that afflict younger people. For example, from 1969 to 1985, the percentage of nursing home residents with a primary diagnosis of a mental disorder rose from 11 to 22% (Hing, 1989, p. 9).

Theoretically, the approach that should be taken in developing risk assessment models for individual conditions within this new category is to

1. describe the fundamental mechanism(s) that causes individual damage events to accumulate (especially the quantitative significance of various contributory factors);
2. elucidate quantitatively the ways in which specific environmental agents enhance the production or prevent the repair of individual damage events; and
3. describe the relationships between the numbers, types, and physical distribution of individual damage events and the loss of biological function or clinical illness.

Unfortunately, no examples are known in which there has been successful quantitative modeling of any of these three types for any chronic cumulative condition. Often, the qualitative nature of individual damage events is not difficult to discern.

For example, atherosclerotic lesions are thought to be produced by a sequence of events described by Ross and Glomset (1976). However, no one has yet successfully developed a predictive model that relates the frequency or severity of these events to the various internal and external causal factors that must be involved. This is clearly an area that requires basic biomedical research, as well as creative interaction between mathematical modelers and experimentalists.

TABLE: Dose-Response Relationship of Parkinson's Disease Risk and Blood or Hair Mercury (Hg) Levels by Approximate Subject Tertiles

Hg Level[a]	No. of Cases	No. of Controls	Adjusted Odds Ratio[b]	95% Confidence Interval
Blood (ng/mL)				
0.0-5.8	6	46	1.0	
5.9-14.2	20	28	8.5	2.2-33.2
? 14.3	28	21	9.4	2.5-35.9
Hair (ppm)				
0.0-4.0	10	42	1.0	
4.1-6.9	15	33	1.8	0.6-5.2
? 7.0	29	20	4.1	1.6-10.5

[a] ng = nanograms; mL = milliliters; ppm = parts per million.
[b] Odds ratio after adjustment by conditional logistic regression for cigarette smoking, alcohol consumption, seafood intake, and ethnic medication.
SOURCE: Ngim and Devathasan (1989).

A couple of consequences of importance here, however, follow from the basic nature of the cumulative irreversible processes that define this category. First, it is important to recognize that the damage processes must be proceeding continually under quite usual everyday situations. This requires viewing with some suspicion the usual presumption of the traditional toxicological paradigm that the common adaptive responses to everyday insults are harmless. For diseases such as atherosclerosis, which proceed silently and chronically in ordinary individuals throughout their lives, homeostatic protective mechanisms must be failing in subtle ways rather frequently. It is likely that there are thresholds that give rise to the small-scale damage events of chronic cardiovascular disease processes [for example, perhaps the lining of the arterial wall in a particular region only suffers appreciable damage when systolic blood pressure temporarily goes above 180 millimeters (mm) of mercury (Hg)]. However, whatever thresholds exist must be low enough to produce a sufficient accumulation of net damage to account for the observation that atherosclerosis and long-term blood pressure increases with age occur in very large numbers of "normal" people.

Second, the basic concept of these diseases as accumulations of many small damage events implies that the number of damage events accumulated by different individuals will show a broad continuous distribution. Because of this, any agent that accelerates the production or prevents the repair of such damage events will shift all, or a very large part, of the exposed population in the direction of worse function. For example, noise exposure will cause people who would otherwise have excellent hearing to have less excellent hearing; people who might have fair hearing without noise can expect to be shifted toward poor hearing, and so forth.

Many years ago, Weiss and Spyker (1974) suggested that methylmercury might accelerate the loss of neurons in adult life and contribute to a "chronic cumulative" process that would fall within this category. Recently, an apparently very sound case control epidemiological study among people in Singapore found a strong association between blood levels of mercury and the risk of Parkinson's disease (Ngim and Devathasan, 1989). Parkinson's disease occurs when a large portion of the neurons responsible for making dopamine in the substantia nigra area of the brain is lost. Thus, if the association with mercury exposure holds and proves causal in later studies, it would be a very important basis on which to quantitatively reevaluate the long-term risks of seafood mercury consumption.

CRITIQUE OF RISK ASSESSMENTS

One of the charges to the committee was to review and critique previous agency efforts to assess the likely human risks of chemical contaminants of aquatic organisms. The committee was asked to suggest directions for the improvement of risk assessment approaches to better serve the needs of decision making on chemical risks, but not to attempt to develop its own set of final risk numbers. The committee has chosen to focus on PCBs and methylmercury as two paradigmatic cases, representing one carcinogen and one noncarcinogen, for which there is a relatively sufficient public record of FDA efforts to assess the risks and benefits of alternative control options.

The PCBs are of special interest in light of the results presented in this chapter. By using conventional carcinogenic risk assessment techniques (although these have considerable difficulties in general and especially as applied to PCBs) and available data on likely average daily intakes in the United States, PCBs are shown to pose by far the largest *potential* carcinogenic risk of any environmental contaminants for which measurements exist. According to EPA's cancer potency factor, the aggregate U.S. lifetime risk could be as high as 6×10^{-5}, corresponding to approximately 120 cancers per year. With FDA's lower estimate of the upper confidence limit (UCL) of the PCB cancer potency, the aggregate risk for the U.S. population would be considerably less, but still not insignificant, 2.7×10^{-6}, implying a maximum toll from commercial seafood of about 5 cancers per year.

These risks would be much greater for the subpopulation consuming relatively large amounts of sport or subsistence caught fish near the FDA tolerance level of 2 parts per million (ppm). For someone who consumes 20 kilograms (kg) per year of 20-ppm fish [which is probably not far from the actual consumption of some izndividuals in some areas of Lake Michigan (Humphrey, 1983a,b)], the upper-confidence-limit lifetime risk that would be expected by using the FDA cancer potency estimates is about 5×10^{-4}; for the EPA cancer potency estimate, the lifetime upper-confidence-limit risk would be somewhat more than 1%. Perhaps in part because of the difference in cancer potency factors, there is also a great difference in the practical policies of the two agencies in regulating PCBs, with FDA generally taking a far less protective posture toward PCB exposure than EPA.[6]

Methylmercury is also of special interest because it shows the operation of the old no-effect level/safety factor paradigm in a practical case where, as it happens, an alternative type of quantitative assessment is also possible based on existing data. In this case, therefore, one can examine what levels of risk might implicitly be accepted by using FDA's rule-of-thumb procedure for dealing with noncancer effects.

In the following material, the committee will assess for both casesthe reasonableness of the calculations that went into FDA's decision making in terms of the information available at the time: andtho improvements that could and should be made by applying more modern analytical techniques to recently available or feasibly obtainable data on population exposures, the nature and mechanisms of potential adverse effects, and likely population dose-response relationships.

To briefly foreshadow the committee's conclusions, three statements can be made:

1. Both assessments (Cordle, 1983; Cordle et al., 1982; FDA, 1979; Tollefson and Cordle, 1986), even at the times they were done, suffered from ad hoc unsystematic approaches to the treatment of population heterogeneity in dosage and (for methylmercury) susceptibility to adverse effects. For mercury, the likely special susceptibility of developing fetuses was mentioned in discussion. However, a tenfold "safety factor" was applied to the lowest blood level reported to produce effects for adults (rather than to a no-effect level of intake, which would have been more consistent with established procedures) in the cited epidemiological studies, without numerical allowance for extra sensitivity of fetuses, without a quantitative dose-response treatment of the data then available for adults to gauge the potential adult risk at the blood level selected as the highest permissible for U.S. consumers, and without a quantitative treatment of the effects of pharmacokinetic differences among people that would tend to make blood

levels and risks as a function of dietary intake more variable than blood levels themselves.

2. Procedures for the systematic quantitative analysis and communication of uncertainties are absent. Both analyses use the "method" of compounding a series of ad hoc allegedly "conservative" assumptions (such as the use of the consumption level of the 90th percentile consumer, the upper 99% confidence limit for the PCB animal potency calculation, but a body weight rather than a surface area scaling rule to translate animal to human dose). The difficulty in picking a single series of point estimates of uncertain parameter values is that after the first few such assumptions have been entered into an analysis, no person on earth can determine where one actually is on an overall distribution of the likelihoods of different outcomes. Monte Carlo simulation procedures are now readily available via personal computer-based software for the calculation of probability density functions of different outcomes, given uncertainties in multiple parameters affecting population risk. Such procedures have been effectively advanced in the context of risk assessment in a recent report by Finkel (1990).
3. Both assessments have been rendered substantially obsolete by the development of recent information related to PCB and to methylmercury risks, changes in the economic conditions assumed in the cost analyses, and possibly changes in residue levels and available options for population exposure reduction.

In the case of PCBs, understanding of mechanisms has advanced to the point where there are now serious proposals for a congener-specific assessment of relative activity (Clement Associates, 1989; Jones, 1988). In the view of the committee, this congener-specific activity analysis must be coupled with a congener-specific treatment of pharmacokinetics, which now seems possible based in part on the human dietary exposure and serum concentration data being collected by Humphrey and coworkers (H. Humphrey, Michigan Department of Public

Health, Lansing, personal communication, 1989). Both congener-specific activity estimates and pharmacokinetics should be used to evaluate not only existing animal carcinogenesis data, but also recently emerging epidemiological information from studies in workers.

In the case of methylmercury, recent data allow preliminary evaluation of the population dose-response relationship for a number of long-term neurological effects from exposure during fetal development (Marsh et al., 1987). The analysis in the following section indicates that the apparent interindividual variability in susceptibility for fetal effects is much broader than that for adult effects. Consequently, although the tenfold safety factor, as applied, appears to offer a reasonable degree of protection for adult effects, projections based on an additional toxicological assumption of lognormal distribution of threshold for effects (logprobit projections) of the fetal dose-response data suggest the possibility of appreciable risk from methylmercury exposure, even at levels to which many people are exposed via the diet. Published projections applying a logistic model to the same data come to similar conclusions (Cox et al., 1989).

Polychlorinated Biphenyls (PCBs)

Background

The PCBs originally came to the attention of FDA through a series of contamination incidents in which PCB-containing oils leaked from electrical equipment directly into food substances, such as rice oil in Japan and Taiwan (the "Yusho" incidents), and by various routes found their way into U.S. livestock. The initial concerns were direct overt toxicity and reproductive effects, and acceptable daily intakes were defined as 100-fold and 10-fold below the no-effect levels in animals and humans, respectively.

To this day, FDA analyses continue to reflect and even emphasize calculations within the NOEL/safety factor paradigm that was initially used to interpret animal and human data on the risk of overt chronic toxicity. Thus, Scheuplein (1988) compares FDA's 1 microgram (μg)/kg/day ADI to the observations of Humphrey (1983a) that average PCB intakes in a group of heavy eaters of fish[8] from Lake Michigan were 1.75 μg/kg/day:

No symptoms or adverse health effects clearly related to PCB ingestion could be identified in the exposed group. This suggests that FDA's acceptable intake level of 1.0 μg/kg-day provides reasonable safety, although it is possible that effects too subtle for detection are occurring or that the latency periods for cancer are very long.

From work reported to EPA (Humphrey, 1983b) and other data reported later by Humphrey 91988), it appears that the cited work had only a very small health component (major emphasis was on defining exposures).

TABLE: FDA Projection of Human Acceptable Levels for PCBs

Study	No-Effect Level (μg/kg body weight/day)[a]	Acceptable Daily Intake (μg/kg body weight/day)[a]	Reference
Subchronic (rats and dogs)	250-300	2.5-3.0	FDA file data
Yusho data (actual 50-day exposure)	200	20	Karatsune and Fukuoka, Acta Med. Med. 23(117), 1971[b]
Yusho data (assuming 1,000-day exposure)	10	1	Fed. Reg. 38(129), July 6, 1973

[a] μg = micrograms.
[b] This is the reference given by Scheuplein (1988). The Toxline data base lists a paper that seems to roughly correspond to Yamaguchi, A., T. Yoshimura, and M. Kuratsune. 1971. A survey on pregnant women having consumed rice oil contaminated with chlorobiphenyls and their babies. Fukuoka, Acta Med. 62(1):117-122 (1971).
SOURCE: Scheuplein (1988).

The health component seems to have been a simple medical history questionnaire administered to the participants, asking them about 18 specific conditions. This limited presentation of the data, without confidence limits or sample sizes, may not do justice to the underlying work. Clearly, however, this kind of study was not designed to be a very sensitive detector of carcinogenic risk or other types of effects (e.g., on fetal development) that are the concerns at low doses. Carcinogenic risks are traditionally studied in human populations by case control studies and detailed comparisons of the frequency of specific cancers or cancer deaths in defined cohorts, adjusted for age.

Reproductive effects must clearly involve careful follow-up of offspring in groups where – through either dietary analysis or serum studies– it is possible to define prenatal exposures. The fact that the Humphrey (1983a) work is seriously cited as evidence of no effect – and that other work (e.g., Fein et al., 1984; Jacobson et al., 1985) that does suggest fetal effects from the same general Michigan fish eater population is not cited – is perhaps testimony to FDA's general lack of enthusiasm at the prospect of departing from the NOEL/safety factor paradigm.

TABLE: Percentage of Humphrey (1983b) Study Participants Reporting Histories of Various Medical Conditions

Condition	Fish Eaters	Comparison Group
Diabetes	4.6	5.9
Heart attack	6.8	3.9
Hypertension	21.1	18.4
Kidney problems	2.2	1.2
Cancer	5.5	4.7
Liver problems	2.4	2.0
Other conditions	11.5	14.3

SOURCE: Humphrey (1988).

FDA Assessment of Costs and Risks for PCBs in Fish

The most recent detailed statements concerning FDA assessment of the carcinogenic risks of PCBs appear to be in two papers by Cordle (1983) and Cordle et al. (1982). These recapitulate and give more details of the analysis underlying the official risk assessment that appeared in the *Federal Register* (FDA, 1979) promulgating the 2-ppm tolerance (lower than the former 5-ppm guideline).

This analysis appears to have been done in part to satisfy requirements for a balancing analysis of benefits and costs that became a government-wide requirement for regulatory action in the mid-1970s. For this purpose, however, the calculation of societal aggregate cancer risks from 90th percentile values of consumption for fish consumers is problematic because, of course, the 90th percentile overstates the average. This probably results in a small error in the opposite direction, because of the likely skewed distribution of PCB consumption among fish consumers.[9] The committee's aggregate national average estimate of current PCB exposure from fish is slightly less than 0.01 μg/kg/day.

Had it done so, FDA might at least have been led to some explicit comment on the closeness (within twofold) of the incremental cost/benefit ratio of the regulatory step it declined to take (from 2 to 1 ppm) to the incremental cost/benefit ratio of the regulatory step it did take (from 5 to 2 ppm). If FDA really intended this result, then an unusually specific insight is afforded into the value it attached implicitly to the prevention of a unit of upper-confidence-limit cancer risk.

Instead, it appears that the FDA decision was most influenced by the concentration of the regulatory impact expected for commercial harvesting of freshwater fish. Whereas the 2-ppm tolerance was expected to cause the loss of a negligible portion of the commercial marine catch (less than 0.2%), the same regulation was expected to cause the loss of 14% of the freshwater catch. For the 1-ppm standard, the loss of freshwater fish was expected to increase to 35%. In its conclusion commenting on the choice of the 2-ppm level, FDA focuses on the absolute magnitude of the losses expected from the 2-ppm and 1-ppm alternatives:

> ... the agency estimates that under a tolerance of 1 ppm, approximately $16 million worth of the commercial fish catch would be violative and thus, presumably, removed from commerce. This is nearly triple the $5.7 million worth estimated to be violative under a 2 ppm tolerance. It is far more likely under a 1 ppm tolerance than under a 2 ppm tolerance that the more heavily contaminated species of freshwater fish would

be violative in percentages high enough to put an end to their commercial exploitation and, possibly, force some segments of the freshwater fishing industry to cease operations completely.

TABLE: FDA Risk and Cost Calculations for PCBs

A. Incremental Health Benefits				
PCB Fish Tolerance Level (ppm)	Heavy Consumers' Projected Dose (μg/day)	UCL[a] Lifetime Individual Risk for Heavy Consumers	Aggregate UCL No. of New Cancers per Year for Heavy Consumers	Incremental[b] Savings of UCL Cancers per Year
5	20.1	9.8×10^{-5}	46.8	
2	14.9	7.2×10^{-5}	34.3	12.5
1	9.1	4.4×10^{-5}	21	13.3

B. Incremental Costs and Cost/Benefit Ratios			
PCB Fish Tolerance Level (ppm)	Projected Cost in Lost Landings (1974 $/yr)	Incremental[c] Cost (1974 $/yr)	Implied Incremental Cost/Benefit Ratio (1974 $/UCL cancer)
5	600,000		
2	5,700,000	5,100,000	410,000
1	16,700,000	10,300,000	770,000

[a] Upper confidence limit. The numbers in this column imply a low-dose cancer potency of about 0.34 cancer case per lifetime average mg/kg/day exposure.

[b] Numbers in this column reflect the reduction in annual UCL-estimated cancers attributable to lowering the standard from 5 to 2 ppm, and from 2 to 1 ppm.

[c] Numbers in this column reflect the increase in annual lost landings cost attributable to lowering the standard from 5 to 2 ppm, and from 2 to 1 ppm.

NOTE: Aggregate projections were apparently made by assuming that all 33 million consumers of the species of interest had intakes at the 90th percentile.

SOURCE: FDA (1979).

Thus, the actual loss of food resulting from the 1 ppm tolerance could greatly exceed even the $16 million landed value (1974 dollars) estimated above. Second, for the average consumer, current exposure to PCBs in fish is at a tolerably low level, when considered in the light of the criteria of section 406 of the act [Food, Drug, and Cosmetic Act], without a 1 ppm tolerance. The average consumer eats a modest amount of fish from a variety of sources, both freshwater and marine, most of which yield fish with PCB levels below 1 ppm. Because their exposure is thus low to begin with, they are adequately protected by a 2 ppm tolerance, which ensures that they will not be exposed to the unusually high levels of PCBs found in some species of fish. The slight additional protection these average consumers might gain from a 1 ppm tolerance does not justify the significantly greater impact such a tolerance would have on the availability of food. On the other hand, atypical heavy consumers (e.g., the Great Lakes sport fisher who catches and consumes large quantities of the contaminated species) would likely not be adequately protected by even a 1 ppm tolerance because of the amount of fish they eat and because those fish are seldom affected by FDA tolerances (either because they are sport fish or are from intrastate commercial channels and, in either case, are outside FDA's jurisdiction).

TABLE: Results of Revising the Aggregate Risk Calculations to Reflect and Assumption that Consumers Had Intakes at the 50th Percentile[a]

A. Incremental Health Benefits				
PCB Fish Tolerance Level (ppm)	Median Consumers' Projected Dose (μg/day)	UCL[b] Lifetime Individual Risk for Heavy Consumers	Aggregate UCL No. of New Cancers per Year for Heavy Consumers	Incremental Savings of UCL Cancers per Year
5	7.57	3.7×10^{-5}	17.6	
2	5.59	2.7×10^{-5}	12.9	4.7
1	3.3	1.6×10^{-5}	7.7	5.2

B. Incremental Costs and Cost/Benefit Ratios			
PCB Fish Tolerance Level (ppm)	Projected Cost in Lost Landings (1974 $/yr)	Incremental Cost (1974 $/yr)	Implied Incremental Cost/Benefit Ratio (1974 $/UCL cancer)
5	600,000		
2	5,700,000	5,100,000	1,100,000
1	16,700,000	10,300,000	2,000,000

[a] This probably understates aggregate risks somewhat.

[b] Upper confidence limit. The numbers in this column are based on a low-dose cancer potency of 0.34 cancer case per lifetime average mg/kg/day exposure.

Some additional portion of the difference derives from the fact that EPA has used more recent experiments with Aroclor

1260 for its risk calculations (Norback and Weltman, 1985), whereas FDA used data on the overall tumor risk from an experiment with Aroclor 1254, which produced the highest estimate of risk from studies then available (Cordle et al., 1982).

In the context of the carcinogenic risk assessment practices of its time, the FDA quantitative analysis does not greatly depart from established principles. On the theory (discussed above) that cancer is the result of a multiple series of somatic mutations, a hypothesis of low-dose linearity in carcinogenic response is not at all unreasonable, particularly if carcinogenic transitions caused by one agent can interact with some of the processes that cause the vast number of "background" cancers in the human population. An upper-confidence-limit risk derived from the multistage theory guarantees low-dose linearity in risk estimates and thus provides a convenient benchmark for risk even though a "mean-estimate."

One can question the practice of summing up tumors from all sites in terms of the multistage theory, but this is not unusual in the context of risk assessment practices in the late 1970s and early 1980s. The choice of animal to human extrapolation formulas is open to considerable discussion in the field to this day, although in the specific case of PCBs, from some limited positive epidemiological studies in PCB-exposed workers, Allen et al. (1987) have estimated a TD_{25} (the lifetime dose estimated to produce cancers in 25% of exposed people) of 0.15 mg/kg/day (for 45-year, 240-day/year occupational exposures). This translates into the equivalent of a best-estimate cancer potency factor of 3.9 cases per lifetime mg/kg/day dosage–rather closer to the EPA upper confidence limit estimate of 7.7 than to the FDA estimate of 0.34 case per lifetime mg/kg/day.

Overall, FDA (1979) has been quite cautious in presenting the difficulties and uncertainties of carcinogenic risk analysis for PCBs from the data it had available:

... the utility of this risk assessment for evaluating actual risk to humans from exposure to PCBs is extremely limited. This is due both to difficulties inherent in making such extrapolations from animals to humans and, perhaps more importantly in this instance, to gaps and uncertainties in the data available for this particular risk assessment. For example, the toxicity studies on which the risk assessment is based used commercial preparations of PCBs, which are chemically different from the PCB residues found in fish and which contain small amounts of highly toxic impurities (e.g., dibenzofurans) not known to be present in fish residues. [It should be noted, however, that these contaminants are also not known *not* to be present in fish.] Also, in making the exposure estimates required for the risk assessment, it was necessary to use existing data on the numerical distribution of PCB levels in fish and rely on the assumption that the effect of a given tolerance level is to remove from commerce all fish containing PCBs exceeding the tolerances. It is possible that neither the assumption nor the data precisely reflect what actually occurs.

For these reasons and others. ... the risk assessment does not provide a basis for precise quantification of the amount of risk reduction accomplished by reducing the fish tolerance. Despite the limitations inherent in the risk assessment, however, the agency regards it as illustrative of the basic validity of the toxicological rationale for reducing the tolerance for PCBs in fish: Reduction of the tolerance will result in a significant reduction in risk among those who consume PCB-contaminated fish. FDA considers this risk reduction to be of significant public health value, even though it cannot be precisely quantified.

For this limited purpose, the analysis as presented may well have served an appropriate and useful function. However, there are now both an opportunity and a need to gather and analyze newer information, so that by the early to mid-1990s it will be possible to significantly improve on this 1979 effort.

Suggestions for Improved Analysis

First, a better regulatory impact analysis might evaluate *a wider range of choices* of regulatory alternatives–including not just different tolerance levels, but possible rules restricting the location of harvest, species, and size. Choices in these areas are likely to allow a more narrowly targeted regulatory action that will reduce human exposure to PCBs at less overall cost of forgone food resources. By being keyed to characteristics of fish that do not require expensive and time-consuming chemical analyses, such regulations might be much more efficiently implementable than tolerances based simply on chemical content.

Second, an improved analysis of carcinogenic risks should be based on a congener-specific assessment of (1) pharmacokinetics and (2) relative potency at the site(s) of action. The FDA analysis mentions the fact that the distribution of PCB congeners to which consumers are exposed is not the same as found in the commercial PCB preparations (e.g., Aroclor 1254) that were the subjects of carcinogenicity testing. The FDA specifically mentions the possibility of dibenzofuran contamination in the commercial mixtures used for carcinogenicity testing. Not mentioned, however, is the likely possibility that fish PCB residues may be relatively enriched in just those congeners that are more persistent in biological systems. More degradable congeners are almost by definition likely to be preferentially destroyed either in the fish themselves or in organisms that are lower on the food chain, whereas the original commercial mixtures would contain the full mix of more- and less-degradable congeners. Humphrey (1983b) noted a change over time in the distribution of PCB congeners found in human serum toward more highly chlorinated congeners, which tend to be more persistent in biological systems. Data now exist that would allow a detailed comparison of congener distributions in fish and humans with those in commercial mixtures (McFarland and Clarke, 1989).

At the same time, a long series of comparative studies by Safe (1989) and others has provided basic insights into the mechanisms of action of PCBs, dibenzofurans, and dioxins. It now appears likely that PCBs enhance carcinogenesis not by direct interaction with DNA, but through a receptor-binding mechanism (similar to dioxins) whose precise dose-response implications have not been elucidated. Receptor binding and related activities vary enormously among different congeners, with congeners that are planar in shape (and relatively highly chlorinated) appearing to have the greatest potency. Careful experiments relating tumor enhancement to the number, type, and persistence of specific PCB-receptor interactions in animal systems could yield important insights into the likely form of the dose-response relationships that should be applied to human risk assessment. There are now serious proposals for a congener-specific assessment of relative activity by using the existing data base (Clement Associates, 1989; Jones, 1988). This approach needs to be pursued and refined with additional laboratory and carcinogenesis studies.

A good opportunity may exist to reevaluate the accumulating evidence of human carcinogenic response in occupationally exposed populations with relation to serum PCB levels and to relate the human response per unit of delivered dose to the response per serum level likely to have been present in rodent bioassays. [There are two very limited but apparently positive epidemiological studies in groups of capacitor workers (Bertazzi et al., 1987; Brown, 1987) and at least one negative study at a plant in Massachusetts whose results have not yet been published in detail. In at least one case, there are measurements of serum PCB levels in the exposed group.] Harold Humphrey of the Michigan Department of Health reportedly has an extensive data set of PCB concentrations in blood in relation to dietary exposure levels. A series of congener-specific measurements is now in progress together with published measurements of metabolic rates of PCBs in humans (Buehler et al., 1988; Phillips et al., 1989; Yakushiji et al., 1984); this is likely to be useful in establishing metabolic rates of PCBs for comparison with animal models and measurements after the work by Matthews and Dedrick (1984) and Tanabe et al. (1981).

Animal work, together with the newer metabolism information, appears capable of providing a basis for estimating serum PCB levels in the rats and mice that showed positive cancer responses in chronic bioassays. Projections to human risk could then be made on the basis of a more appropriate measure of "delivered dose" (weighted for different congeners by receptor binding and pharmacokinetics) rather than gross dietary intake. There already is reasonably good information on the population distribution of overall plasma PCB levels in the United States Using such a distribution, if one had an estimate of cancer risk per unit of plasma PCB concentration, an inference could be made not only about the average risk to the population, but about the distribution of risk to various individuals that derives from differences in both dietary habits and PCB elimination rates. Plasma and whole body burden PCB levels could also be productively used as indicators of long-term dosage in epidemiological case control studies of cancer patients (versus other people) in communities near the Great Lakes with greater than usual numbers of people having high exposures to fish with appreciable levels of PCBs.

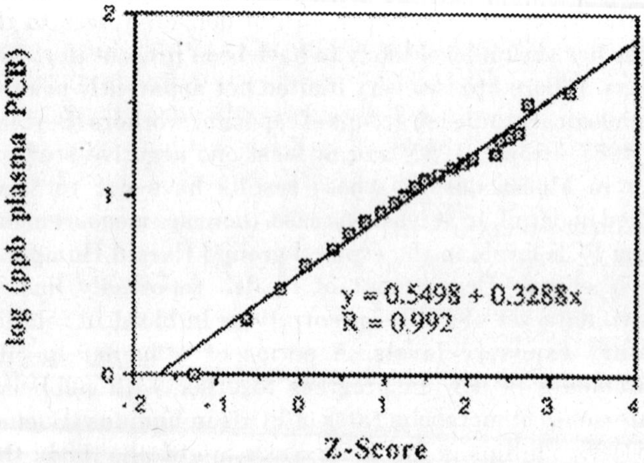

FIGURE: Plasma PCB levels in 738 Southern Californians screened prior to employment [data of Sahl et al. (1985)]

A better analysis should also include a quantitative assessment of the risks posed by prenatal exposure to PCBs, via such parameters as gestational period and birth weight, as well as subtle neurological measures of effect. The presence of at least modest effects on birth weight has recently been supported by an epidemiological study in workers (Taylor et al., 1989). Because birth weight is such an easy parameter to study (measurements are made routinely on nearly all newborns) and because it has a very strong relationship to infant mortality (Hogue et al., 1987), additional large population studies of birth weight and gestational period in relation to fish consumption and relevant serum PCB congener levels should be a high priority. Methodology exists to interpret birth weight changes in terms of other relevant endpoints such as infant mortality (Ballew and Hattis, 1989), but these techniques require validation with an actual case having large enough numbers of subjects to provide data on both the intermediate parameters (birth weight and gestational age) and the ultimate parameters of concern (infant mortality, impairment of neurological development, etc.).

Finally, there is a need for improved and more systematic treatment of analytical uncertainties, with ultimate results in the form of cumulative probability density functions or, failing this, at least a statement of both best-estimate and upper-confidence-limit risks, under consistent and defensible theories.

Methylmercury

FDA Risk Assessment and Current Regulations

The published analysis of Tollefson and Cordle (1986) appears to be the most definitive public assessment by FDA personnel of the risks of methylmercury in fish.

The principal benchmark that Tollefson and Cordle (1986) use in defining an acceptable intake of methylmercury is the lowest blood methylmercury level that appeared to be associated with overt neurological symptoms of toxicity in a few studied mass poisoning incidents (two that occurred from fish taken from Minamata Bay and Niigata in Japan in the 1950s and 1960s, and one incident that arose from the consumption of contaminated seed grain in Iraq): A Swedish expert group evaluated data on human methylmercury toxicity derived from cases which occurred in Minamata and Niigata. The Swedish group determined (by extrapolation) that the lowest blood mercury level associated with toxic effects was 220 ppb [parts per billion] and the lowest hair mercury level associated with toxic effects was 50 ppm (Berglund et al., 1971).

The Iraqi outbreak of methylmercury poisoning has been extensively studied by many investigators. The blood level of mercury at which symptoms of toxicity were first detected in the Iraqi episode was approximately 240 ppb [parts per billion] (Clarkson et al., 1976). This calculation was made on samples collected 65 days after the end of exposure, which is the approximate half-life of methylmercury in humans. Since the actual clearance times from the blood are not known, the level may lie between 240 and 480 ppb. These values are for adult exposures only.

Studies of people in Sweden who consumed relatively large amounts of fish allowed these blood levels to be related to dietary levels at least on a population average basis (Tollefson and Cordle, 1986): "A linear relationship was found between daily ingested methylmercury and the level of methylmercury in blood, and the data indicated that a steady daily intake of approximately 300 μg Hg [mercury] as methylmercury for a 70-kg person would result in a blood concentration of roughly 200 ppb at steady state (Skerfving, 1974)."

This 300-μg/day dose rate, calculated to correspond to 200 ppb in blood *for an average person*, was interpreted by Tollefson and Cordle (1986) as a threshold value at which symptoms of toxicity associated with methylmercury are first noticeable. ... Dose-response relationships below this range of intake are not known.

In addition, there is concern about the relative sensitivity of the developing fetus. The question of interaction of other chemical factors such as selenium on the toxicity of methylmercury has not been conclusively demonstrated at this time, but may be a factor to be considered when more information is developed. Because of these areas of uncertainty, a safety factor of ten has been used to provide a sufficient margin of safety.

Thus a maximum tolerable level would be 30 μg methylmercury daily in the diet, resulting in 20 ppb of methylmercury in blood and 5 ppm in hair.

The following limitations to this approach were recognized:

(1) it was not known to what extent particular individuals are more or less sensitive to mercury than others;

(2) the estimates were based on the "lowest level that caused an effect" rather than the normal procedure of using a "no-effect dose level";

(3) paresthesia is usually the first symptom of methylmercury toxicity noted but is not sufficient to diagnose poisoning because it can be caused by many other factors (Clarkson et al., 1983);

(4) questions about dose-response relationships in human fetuses and newborn infants were unanswered; and

(5) there is a possibility of subclinical effects arising from exposure to very low levels of methylmercury.

Tollefson and Cordle (1986) therefore realize that they are departing from the usual procedure for the NOEL/safety factor analysis and that, in particular, risks to fetuses may not be represented adequately in the available data. However, they make no additional numerical adjustment for these uncertainties, presuming implicitly that the usual 10-fold safety factor has enough of a margin built into it to avoid an unacceptable level of risk.

In addition to these enumerated factors, there is a subtle point that makes the calculation on the basis of the average diet/blood relationship less protective than the usual calculation based on a 10-fold reduction from the dietary dose that is directly observed to be without effects.

The 10-fold factor is supposed to represent the protection needed to guard against interindividual variability in the complete pathway from dietary intake to the production of biological responses. Some portion of that human interindividual variability, however, occurs in characteristics that affect individual blood levels—especially individual half-lives for elimination of methylmercury from the body. For those with relatively slow elimination rates, toxicity would be expected to be associated with lower average dietary intakes than for those with average elimination rates. Therefore, by doing the dose extrapolation in terms of blood levels, and translating the corresponding *average* dietary levels, the authors are being less protective than would have been the case had the 10-fold factor been applied to the minimum dietary dosage associated with an observed effect (or no observed effect). Later in the paper, Tollefson and Cordle (1986) discuss interindividual differences in biological half-lives for methylmercury as follows:

Additional data developed on the biological half-life of methylmercury in humans, however, indicate a need to take into account the problem of variations among individuals. In the Iraqi episode, 90% of the individuals studied had a biological half-life of methylmercury between 60 and 70 days, but 10% showed values of 110 to 120 days (Al-Shahristani and Shihab, 1974). Individuals having a long biological half-life would accumulate much higher steady-state levels than those having short biological half-lives and would thus be at greater risk from the same level of methylmercury intake.

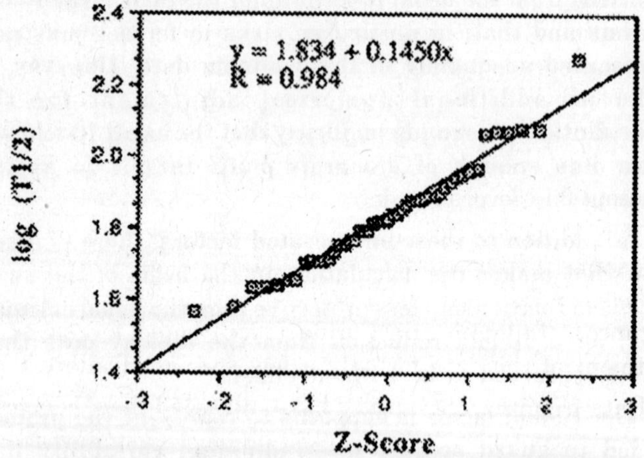

FIGURE: Lognormal plot of the distribution of methylbercury biological half-lives (data of Al-Shahristani and Shihab, 1974)

Tollefson and Cordle have unfortunately misread the population distribution of methylmercury half-lives presented in the cited paper. A figure in the paper is misleading in that the highest single observed biological half-life (189 days) has somehow been excluded, but there is no way that even the figure can be construed to indicate that 90% of the values are between 60 and 70 days.

Tollefson and Cordle (1986) go on to support the results of their analysis by citing studies failing to find obvious methylmercury toxicity in heavy fish-eating population groups in Peru and American Samoa with average blood methylmercury concentrations in the range of 60-80 ppb (Marsh et al., 1974; Turner et al., 1980). However, without careful analysis of the distributions of thresholds implied by the earlier data, the sample sizes; and the degree of background symptoms in control groups in the negative studies cited, neither the reader nor the authors can judge the sensitivity of those studies to detect adverse effects and whatever quantitative implications there might be for the level of risk at lower doses.

Tollefson and Cordle's apparent reluctance to construct and analyze the implications of quantitative population distributions is also seen in their treatment of the expected dietary dosage of methylmercury to consumers. They had available the results of a fairly extensive study commissioned by the Tuna Research Foundation (TRF) of one-month diary reports of consumption of different types of fish by 7,662 families, including 25,165 people, and measurements of mercury levels in different species of fish. However, instead of using these data to describe the population distribution of monthly mercury intakes from fish as a whole and the possible effects of different tolerances, Tollefson and Cordle (1986) opted for an ad hoc treatment of the data for only a few species of interest:

> Data from the TRF survey have indicated that the average consumption of all fish among the fish-eating population of the United States is 18.58 oz/month or approximately 18 g/day. Daily consumption of species containing relatively high levels of methylmercury such as tuna, swordfish, or halibut would be considerably less. For example, the mean daily consumption of halibut is 7.2 g with two standard deviations increasing the total to 16.6 g. This would include some 97.5% of all consumers of halibut. The consumption of 16.6 g of halibut with 0.179 ppm mercury would provide a daily mercury intake of approximately 2.9 μg.

The mean daily consumption of swordfish is 6.53 g with a standard deviation of 2.5 g. If 11.53 g of swordfish with a mercury level of 1.5 ppm were consumed each day, and this would include over 95% of all swordfish eaters, the daily mercury intake would be 17.3 μg, still below the ADI of 30 μg.

At the highest level of swordfish consumption shown by the NPD [National Purchase Dairy Panel, Inc.] study, i.e. 511 g in a month or 17 g per day, with a mercury residue of 1.5 ppm the daily mercury intake would be 25.5 μg, still below the ADI of 30 μg. If in addition to the highest level of consumption of swordfish in the NMFS [National Marine Fisheries Service] study, i.e. 17 g per day with a mercury residue of 1.5 ppm, the same individual consumed the average daily amounts of tuna, halibut, and salmon at the present average residue levels of mercury for each of the three species, then the total daily intake of mercury from all four species would average 31.7 μg, only 1.7 μg above the ADI. Such consumption, which would realistically be on a periodic basis to derive daily exposure, seems very unlikely, particularly since the cost of swordfish, halibut, and salmon is prohibitively high.

Tollefson and Cordle are certainly correct in concluding that very few consumers will exceed the 30-μg/day level for prolonged periods. However, this type of analysis cannot give an estimate of how many might, or how many might fall short of the 30-μg/day level by various amounts. The failure to analyze the data more rigorously is all the more unfortunate because the FDA had available an example providing statistically sophisticated treatment of these same data (NMFS, 1978).

Tollefson and Cordle (1986) conclude with a somewhat unusual statement of the relevant risk issues:

The data currently available for evaluating quantitatively the association of neurological symptoms of toxicity with exposure to methylmercury are sparse and inconsistent. Additional studies are being carried out on the prenatal effects of methylmercury to determine that this lifestage continues to be protected by the 1.0 ppm regulatory level for mercury in fish. However, even with the above-outlined uncertainties concerning the results of exposure in Japan and Iraq, where exposures were considerably higher than anything experienced in other countries, U.S. fish consumption data do not indicate any cause for concern of methylmercury poisoning for the *average American.*

The majority of fish consumers could easily double their intake and still remain below the mercury ADI. The current 1.0 ppm regulatory level for marine species provides more than adequate protection *at the current average fish consumption level in the U.S.* In addition, the enforced limit of 1.0 ppm mercury in marine fish provides a sufficient margin of safety for young children and for significant numbers of consumers exceeding the acceptable daily intake.

The potential for a more quantitative analysis of methylmercury risks with the aid of available data on the distributions of blood methylmercury concentrations and dose-response data from the Iraqi poisoning incident is illustrated below. The aim is not to provide a definitive assessment of methylmercury risks, but to show an approach for improving the type of assessment now done by FDA.

Available Data on Blood Levels in the U.S. Population

Gathering and analyzing data on the actual distribution of blood mercury and methylmercury levels in the population allow one to bypass a great deal of the complexity of modeling individual dietary intakes over times long enough to determine biologically significant systemic exposures. Full data describing this distribution are not available; however, a plot of the median and extreme values given indicates that the distribution is well described as lognormal.

The variability indicated by the $\log_{10}$ geometric standard deviation of 0.328 indicates that 95% of the population is spread out over a 20-fold range – between 4.5-fold above and 4.5-fold below the median blood level for the population. This variability includes both dietary and pharmacokinetic differences among individuals.

Algonac (population 3,684) was chosen because of its proximity to both Lake St. Clair and the St. Clair River – sites of a famous mercury contamination episode discovered by a

Canadian graduate student in 1970. South Haven (population 6,471) was chosen as a control community with similar proximity to Lake Michigan. Recently, Marsh et al. (1987) published detailed information on the incidence of a variety of fetal mercury effects in relation to the maximal levels of mercury found in the hair of the mothers during gestation. (The observations come from an Iraqi mass poisoning incident that resulted from the distribution of methylmercury-treated Green Revolution seed grain.) In all, 81 mother-infant pairs were studied.

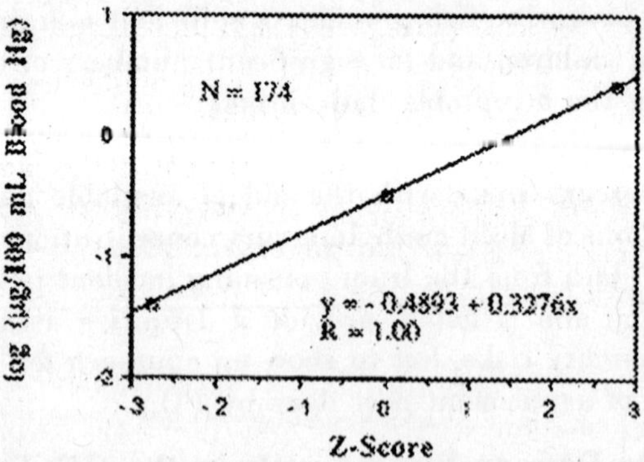

FIGURE: Lognormal plot of the Haxton et al. (1979) data on blood mercury concentrations (only the median and extreme values of the distribution were available for plotting)

Such a conclusion, however, depends on whether the biomarker of exposure used in this case – the maximum hair mercury found at any time during gestation – is the most appropriate direct causal predictor of response that can be developed. Other possibilities might well include the concentration of mercury at a specific sensitive time during gestation or a weighted sum of concentration x duration over a specific set of sensitive periods. Accurate assessment of the degree of interindividual variability in susceptibility in humans, and consequent low-dose risks, might well depend on quantitative measurement and modeling of the causal processes involved in this case and reanalysis of the data according to the most likely causally predictive summary measure of delivered dose.

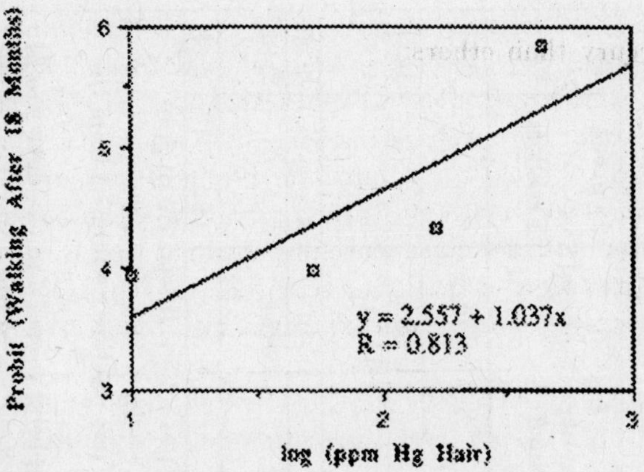

FIGURE: Logprobit dose-relationship for walking after 18 months in relation to maternal hair mercury (data of Marsh et al., 1987)

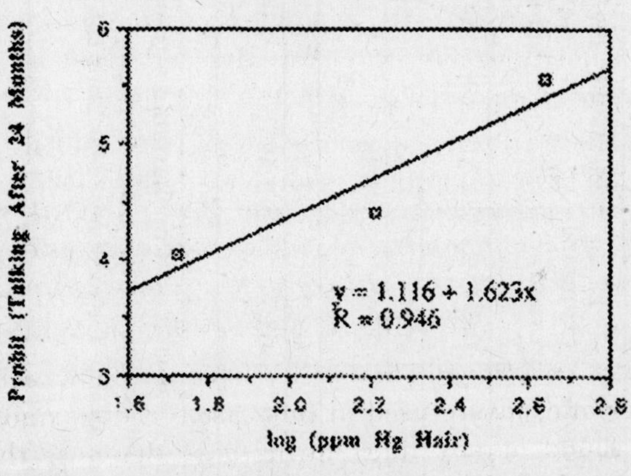

FIGURE: Logprobit dose-response relationship for talking after 24 months in relation to maternal hair mercury (data of Marsh et al., 1987)

It can be seen that, as with fetal effects, the fuller analysis of data with the Finney (1971) procedure suggests somewhat larger probit slopes than the values determined by simple regression analysis of the points. It can also be seen that, in comparison with the fetal effects, the effects in adults tend to have larger probit slopes, indicating less interindividual variability and therefore less risk at low doses.

FIGURE : Logprobit dose-response relationship for "mental symptoms" in children in relation to maternal hair mercury (data of Marsh et al., 1987)

Effect	Background Response (%[b])	Probit Slope	Slope Standard Error	Intercept	ED_{50} (ppb blood)	ED_{50} Geometric Standard Error	Chi-Squared	Degrees of Freedom[c]	P[d]
Late walking	0	1.21	0.30	2.19	205	1.49	6.093	3	0.11
Late talking	7.3	1.76	0.71	0.81	244	1.38	0.689	1	0.41
Mental symptoms	2.4	0.99	0.76	1.88	1,429	4.75	0.351	1	0.55
Seizures	0	1.10	0.53	1.54	1.399	2.95	0.356	3	0.95
Neurological score >4	0	0.85	0.27	2.42	1.047	2.54	0.874	3	0.83
	Average	1.18				Sum	8.363	11	0.68

[a] The equation fit is: probit of excess risk over background = intercept + (slope) × $\log_{10}$ (blood Hg in ppb).

[b] Estimated from data in the lowest one to three dose groups.

[c] The number of dose groups available for analysis, less 2 for the number of parameters estimated from the data (intercept and probit slope).

[d] The probability that a deviation as large as that observed between the logprobit model and the data would have been expected by chance, even if the logprobit model were a perfect description of the underlying dose-response function.

Biologically, the implication is that the more complicated array of causal processes involved in producing fetal effects (allowing for variability in maternal mercury elimination, maternal/fetal transfer, and the time-sensitive processes of interference by methylmercury with development) tends to lead to more interindividual variability. From a social policy perspective, the implication is that the 10-fold safety factor rule, whatever its effectiveness in protecting exposed populations against the effects of interindividual variability in adults, appears likely to provide less protection for the developing organism.

TABLE: Logprobit Projections of Risks of Adult and Fetal Methylmercury Effects, Based on Observations from Iraq of Effects as a Function of Blood and Hair Levels (After a Twofold Upward Adjustment of Observed Blood Levels Because of Measurement Delay)

			Blood Levels Required for Different Degrees of Risk			
Effects	Hair ED_{50} (ppm)	Blood ED_{50} (ppb)	ED_{10} (1.2816 SDs below ED_{50})	ED_{01} (2.3263 SDs below ED_{50})	$ED_{1/10,000}$ (3.719 SDs)	$ED_{1/million}$ (4.7536 SDs)
Adult						
Paresthesias		1,145	294	97	43	22
Ataxias		1,687	795	431	275	190
Visual changes		2,010	522	174	7[illegible]	40
Disarthria		2,950	1,569	937	643	472
Hearing defects		3,880	2,449	1,684	1,280	1,022
Deaths		5,010	3,390	2,470	1,959	1,618
Fetal						
Late walking	205	819	72	9.9	0.71	0.10
Late talking	244	978	182	46.3	7.44	1.92
Mental symptoms	1,429	5,720	289	25.3	0.99	0.09
Seizures	1,399	5,600	383	43.0	2.33	0.27
Neurological score >4	1,047	4,190	132	7.9	0.18	0.01

NOTE: Hair ED_{50} equivalents are not given for the adult effects because the primary measurements were made in blood. Because the underlying blood mercury measurements for adults were made after the end of exposure, it is likely that they understate actual blood levels during exposure by about twofold; therefore, the blood levels associated with all the stated levels of risk should also be adjusted upward by about twofold.

Projections of fetal risks here, of course, are greatly affected by the assumed lognormal form of the distribution of thresholds for both adult and fetal effects. Although lognormal distributions appear to be broadly compatible with the high-dose data (see earlier figures), this is not the only statistical form that the distribution of thresholds might take. Cox et al. (1989) have applied a logit model, which assumes a different distribution of thresholds, to predict possible low-dose risks for one of the fetal effects covered here (late walking), using the same data set (Cox and coworkers also made projections based on scores for neurological impairment, but using a different cutoff for defining effect than the committee, thus rendering the risk projections noncomparable).

TABLE : Logprobit Projections of Risks of Adult and Fetal Methylmercury Effects, Based on Observations from Iraq of Effects as a Function of Blood and Hair Levels (After a Twofold Upward Adjustment of Observed Blood Levels Because of Measurement Delay)

			Blood Levels Required for Different Degrees of Risk			
Effects	Hair ED_{50} (ppm)	Blood ED_{50} (ppb)	ED_{10} (1.2816 SDs below ED_{50})	ED_{01} (2.3263 SDs below ED_{50})	$ED_{1/10,000}$ (3.719 SDs)	$ED_{1/million}$ (4.7536 SDs)
Adult						
Paresthesias		2,290	588	194	44	15
Ataxias		3,370	1,591	862	381	208
Visual changes		4,010	1,044	348	81	27
Disarthria		5,900	3,140	1,875	944	566
Hearing defects		7,750	4,900	3,370	2,040	1,411
Deaths		10,020	6,790	4,940	3,240	2,360
Fetal						
Late walking	205	819	72	9.9	0.71	0.10
Late talking	244	978	182	46.3	7.44	1.92
Mental symptoms	1,429	5,720	289	25.3	.99	0.09
Seizures	1,399	5,600	383	43.0	2.33	0.27
Neurological score >4	1,047	4,190	132	7.9	0.18	0.01

NOTE: Hair ED_{50} equivalents are not given for the adult effects because the primary measurements were made in blood. Because the underlying blood mercury measurements for adults were made after the end of exposure, it is likely that they understate actual blood levels during exposure by about twofold; therefore, the blood levels associated with all the stated levels of risk should also be adjusted upward by about twofold.

As mentioned earlier, it would be more common for a 10-fold safety factor to be applied to the dietary dosage of a toxicant, rather than to blood levels. To represent this for methylmercury, one must add to the logprobit risk equations an estimate of the interindividual variability in internal blood levels as a function of dietary intake. In principle, two factors should be considered in accomplishing this: (1) interindividual variability in absorption, and (2) interindividual variability in overall elimination of mercury from the blood. Methylmercury is relatively efficiently absorbed from food, so the influence of the first factor may well be small.

TABLE: Projected Risks at the Presumed Adult Blood Methylmercury LOEL and at One-Tenth the Presumed Adult LOEL

Effects	Probit Slope	Blood ED_{50} (ppb)	SDs Below ED_{50} at Presumed Adult LOEL (200 ppb in blood)	Risk (expected fraction affected) at Presumed Adult LOEL	SDs Below ED_{50} at 1/10th Adult LOEL (20 ppb in blood)	Risk (expected fraction affected) at 1/10th Adult LOEL
Adult						
Paresthesias	2.170	1,145	2.64	0.050	3.81	6.8×10^{-5}
Ataxias	3.926	1,687	3.64	1.4×10^{-5}	7.56	Very small
Visual changes	2.192	2,010	2.19	0.014	4.39	5.8×10^{-6}
Disarthria	4.67	2,950	5.46	2.4×10^{-8}	10.13	Very small
Hearing defects	6.423	3,890	8.27	Very small	14.69	Very small
Deaths	7.582	5,010	10.60	Very small	18.19	Very small
Fetal						
Late walking	1.214	819	0.74	0.229	1.96	0.0251
Late talking	1.756	978	1.21	0.113	2.97	0.0015
Mental symptoms	0.988	5,720	1.44	0.075	2.43	0.0076
Seizures	1.100	5,600	1.59	0.056	2.69	0.0035
Neurological score >4	0.854	4,190	1.13	0.130	1.98	0.0237

NOTE: Because the underlying blood mercury measurements for adults were made after the end of exposure, it is likely that they understate actual blood levels during exposure by about twofold; therefore, the blood levels associated with all the stated levels of risk should also be adjusted upward by about twofold.

The effect of this is to increase the estimate of the probit slope by a relatively greater amount for those effects that, in the absence of the adjustment, had relatively small amounts of interindividual variability (high probit slopes). The corresponding estimates of adult risks are increased, but generally not by a huge factor, for the adult effects expected to pose the numerically greatest risks.

Conclusions and Recommendations for Changes in Risk Assessment Practices

As indicated even by these two examples (PCBs and methylmercury), where FDA has devoted much more attention to two types of residues than to other potential chemical hazards of seafood consumption, the agency lags badly in the development of innovative methodology for assessing risks and evaluating the potential benefits of different options for control. This is all the more unfortunate because FDA has major technical resources available in both risk assessment and experimental toxicology (including reproductive toxicology) at the National Center for Toxicological Research in Arkansas. Neither the PCB nor the methylmercury analyses by FDA show evidence of input from this very capable group, which is under FDA's jurisdiction.

Carcinogenesis risk assessment procedures should be modified to give decision makers additional information about the uncertainties of the analysis, as well as both societal aggregate and individual risk estimates. In particular, to facilitate evaluation of the costs and benefits of measures to achieve quantitative reductions in exposure to carcinogenic inadvertent contaminants, procedures should be developed to supplement current "upper-confidence-limit" cancer potency estimates with estimates representing the central tendency of cancer risks, information on cancer risks from all available species, and comparative information on the pharmacokinetic and pharmacodynamic factors in different species.

The FDA is most conspicuously backward in the development of quantitative risk assessment approaches for noncarcinogens. The rule-of-thumb (ADI/NOEL/safety factor) procedure now universally in use has serious conceptual flaws, inappropriately mixes technical and social policy presumptions in analysis, and fails to encourage the development of better information on pharmacokinetics, human interindividual variability, and other topics that could allow better estimates to be made of human risk for noncancer effects.

ESTIMATING HUMAN INTAKE OF CONTAMINANTS FROM SEAFOOD AND SOME ASSOCIATED RISKS

In this section, the committee pursues a couple of different strategies for estimating exposures and, in some cases, risks for the array of other seafood hazards it was unable to investigate as deeply as PCBs and methylmercury. After available data bases for assessing exposures are reviewed briefly, national average intakes are estimated for various inorganic contaminants by using the FDA Total Diet Study and the Hall et al. (1978) data base, and making comparisons with regulatory levels for inorganics (similar comparisons are not presented for organic carcinogens, because the threshold assumptions implicit in the ADI methodology are not appropriate for the major effects of concern in these cases). After this, national average exposures for both organic and inorganic contaminants in commercial seafood are estimated tentatively from FDA surveillance data and, with the aid of EPA upper-confidence-limit estimates, the potential national cancer risk from organic carcinogenic contaminants is estimated. Finally, the section on exposure from sport, subsistence, and tribal fishing, and the section on the potential for reducing exposure through consumer information or labeling programs and fishing advisories, provide comments on the potential helpfulness of consumer and fisher advisories in reducing exposures.

Review of Data Bases Available for Estimating Exposures

FDA Total Diet Study

The FDA Total Diet Study provides a direct estimate of the dietary pesticide intake from a variety of foods including a limited number of seafood products (Gunderson, 1988). In the Total Diet Study, 234 different food items representing the diets of U.S. consumers are collected and analyzed four times each year throughout the United States. Each of four market basket samples is a composite of foods collected in three cities in a particular region (Gunderson, 1988). The foods are prepared table-ready and then analyzed for residues of industrial chemicals, pesticides, and metals such as cadmium and lead. The principal objective of the Total Diet Study is to develop dietary intake information on industrial chemicals and pesticides and to compare these intakes with acceptable daily intakes (ADIs). An ADI is the daily intake of a chemical which, if ingested over a lifetime, appears, according to FDA, to be without appreciable risk. The ADIs are established by scientific experts who have attended annual joint meetings of the United Nations Food and Agriculture Organization (FAO) and the World Health Organization (WHO) (Gunderson, 1988). The Total Diet Study could be a useful data base for determining the proportionate contributions of chemical and pesticide contamination of seafood and the contributions seafood makes to contamination, relative to other foods such as meat, poultry, dairy products, fruit, vegetables, grains, and oils. However, the FDA Total Diet Study included very few seafood samples among the 234 food items analyzed, and those items that were sampled – tuna, haddock, pollock, shrimp, and fish sticks – were among the products least likely to harbor significant chemical or pesticide contamination. Nonetheless, the pesticides detected in these products cast doubt on the ability of the FDA regulatory monitoring program to detect all pesticides that may be in seafood products (Gunderson, 1988). For example, pesticides such as malathion, diazinon, penta, and chlorpropham detected in seafood products in the Total Diet Study were rarely, if ever, detected by the FDA regulatory monitoring program (Gunderson, 1988). Data from the Total Diet Study are presented later in this chapter.

FDA Pesticide Monitoring Files

The committee's use of FDA pesticide monitoring files is discussed at length in exposures from commercial seafood. Although there are appreciable concerns about the extent and representativeness of these data for specific species and areas of the country, overall they are the best source of information on organic contaminants in the commercial seafood market. However, information on inorganic contaminants in fish muscle is decidedly lacking, and the committee believes that FDA should consider bolstering its data base with measurements on edible portions of finfish for inorganic chemicals such as arsenic, cadmium, lead, and selenium.

The FDA seafood industrial pollutant and pesticide monitoring program has a tremendous potential to deliver important data about the contaminant levels in interstate seafood intended for human consumption. The committee acquired these files, numbering some 1,000 pages, for the years 1985-1988. They are rich in data, and fairly well organized and structured; yet they can also be quite superficial, with significant data gaps.

Although the full files for the most recent and the earliest years were not acquired by the committee, it is believed that complete files were acquired for 1986 and 1987. These files are among the most important sources of information about the concentration of industrial pollutants and pesticides in raw edible portions of finfish and shellfish intended to be sold in the nation's commercial markets. This monitoring effort stresses sampling of fish at the wholesale market level and is not generally useful for identifying the site of the catch and thus the exact source of the residues. However, the files are somewhat species specific. The program is designed primarily to satisfy FDA's legislative mandate of regulating shipments in interstate commerce. The fiscal year 1988 domestic monitoring program requires that a minimum of eight samples of locally produced fish or shellfish of commercial significance be collected per district as close to their origin as possible. Species selected are said to be those with the highest potential for residue contamination. In most cases, nonmigratory bottom feeders are the species of choice for pesticides and PCBs. Fresh fish are collected whenever possible. In addition to the formal domestic program requirements, several FDA district offices have ongoing cooperative programs in which samples from specific areas are submitted to a particular district laboratory for analysis.

The FDA monitoring program is divided into compliance and surveillance samples. The compliance samples are of food items the agency believes to be of special concern (i.e., to pose regulatory compliance problems). These include methylmercury in swordfish or the many freshwater fish that may contain excessive levels of pesticides and industrial chemicals such as PCBs and 2,3,7,8-tetrachlorodibenzo-*p*-dioxin (TCDD). The surveillance samples are representative of normal interstate seafood items for which no compliance problems are expected.

However, it is important to note that data on 80-90% of the fish selected for the committee's work were derived from FDA surveillance sampling. In only a few cases, such as tuna, swordfish, and shark, were samples listed under the compliance program. Thus, in the committee's estimation, the data should not be construed as being too heavily skewed toward the most elevated toxic concentrations. In calculations made from these data, residue levels below detection limits were considered to be zero. Because of this, in the committee's opinion, calculations based on FDA data are likely to underestimate the consumer's daily exposure to industrial chemicals, pesticides, and inorganic contaminants.

The FDA claims that the import monitoring component of this program stresses freshwater species, because strictly deep-sea species are not considered by FDA to be a significant source of pesticide residues. (However, the committee has noted that several deep-sea species such as swordfish and sablefish may indeed be sources of significant pesticide and inorganic contaminant residues.) The FDA does sample many imported saltwater species.

In fiscal year 1984, 453 samples were collected and analyzed (374 domestic and 79 imported samples), whereas in fiscal year 1985, 532 samples were analyzed (464 domestic and 68 imported). The fiscal year 1985-1986 reports appear to include a much larger number of domestic seafood samples; however these data include approximately 900 bluefish samples analyzed to fulfill the 1985-1986 congressionally mandated Atlantic Coast bluefish survey. There are basic problems with FDA data. Among them are high detection limits for chemicals of concern, lack of monitoring for industrial pollutants such as PCBs and pesticides in interstate-trafficked bivalves, and lack of monitoring for inorganic contaminants such as arsenic, cadmium, and lead in edible portions of finfish.

National Marine Fisheries Service Survey

In 1978, the NMFS published a 300-page survey of trace elements in the fishery resource (Hall et al., 1978). This document reports trace element levels in tissues of 204 species of finfish, molluscs, and crustaceans from 198 sites around the coastal United States, including Alaska and Hawaii. Fifteen elements were determined: antimony (Sb), arsenic (As), cadmium (Cd), chromium (Cr), copper (Cu), lead (Pb), manganese (Mn), mercury (Hg), molybdenum (Mo), nickel (Ni), selenium (Se), silver (Ag), tin (Sn), vanadium (Va), and zinc (Zn). The species analyzed represented approximately 93% of the volume of the U.S. commercial and sport fish catch. It is an invaluable, well-conducted study.

Fisheries of the United States, 1987

The *Fisheries of the United States, 1987* report (NMFS, 1988) details the amount of important commercial seafood species that were landed in 1987 by U.S. vessels and imported into the United States from other nations. The committee relied upon this report to estimate per capita consumption figures of important commercially caught seafood.

Exposures from Sport, Subsistence, and Tribal Fishing

Regional reports provided the principal available information for examining industrial chemical and pesticide concentrations in certain species of recreational/subsistence caught seafood and for estimating the intake of such species by anglers and their families. The committee relied on reports from Quincy Bay, Mass.; New York; southern California; and Puget Sound, Wash. These data, however, were not extensive enough to allow national estimates of exposure to be developed via this route.

Exposures from Commercial Seafood

Dietary Intake of Seafood of Various Species

Perhaps one of the most crucial tasks that the committee faced in determining the efficacy of present programs to protect the public health with regard to industrial chemicals and pesticides was determination of the daily dietary intake of specific commercial seafood species. There were several sources to which the committee had access to learn about the daily per capita intake of certain seafood species.

The NMFS *Fisheries of the United States, 1987* report was the most important resource for estimating the intake of many individual commercial seafood species (NMFS, 1988). Unfortunately, trade associations, in general, have not conducted recent consumption surveys (Konz, 1988). The committee used several regional portraits to create consumption, daily exposure, and cancer risk scenarios for typical sport fishers.

Enough information was available for the committee to make regional and national projections of seafood consumption for both commercially sold seafood and sport harvested species. In the future, because each region has different consumption patterns with different pollution problems, a variety of data sources from specific areas will play an important role in aiding risk managers to protect particular consumers, especially regional sport fishing subgroups that may concentrate their intake on certain favored but contaminated species.

The nationwide daily per capita intake of domestic and imported seafood was estimated from landing information contained in *Fisheries of the United States, 1987;* landings data for each species were given in round (live) weight, that is, the weight of the catch as aquatic animals are taken from the water before any processing (NMFS, 1988). After subtracting exports of whole uneviscerated products and annual changes in inventories (when available), the edible quantity of each species was calculated with the aid of a conversion factor supplied by NMFS. Generally, these conversion figures ranged from 0.35 to 0.59% of the live round weight. Upon application of the conversion figures, exports were subtracted under the product categories of fillets, steaks, portions, and cured products (NMFS, 1988).

Shellfish landings for univalves and bivalve molluscs, such as clams, oysters, and scallops, are reported in weight of meat, excluding the shell (NMFS, 1988, p. 2). Conversion factors were, however, applied to crabs, shrimp, squid, and the miscellaneous category of other marine and freshwater shellfish.

After applying the appropriate conversion factors and subtracting exports and year-end inventory, the amount of product remaining was divided by 242 million, the approximate U.S. population in 1987. This figure, which provided annual per capita consumption data, was then divided by 365 days to determine daily per capita consumption.

For imports, conversion factors were applied when whole or eviscerated fish were listed. Other imports were listed under fillets and steaks; canned products; balls, cakes, and puddings; and pickled or salted product; these categories, which dealt with edible portions, required no conversions. The amount imported – after the necessary conversions – was again divided by 242 million (the approximate U.S. population in 1987) and then by 365 days.

Unfortunately, the data provided for imports frequently neglected individual aquatic species. Thus, it was not possible to determine per capita consumption for certain aquatic species such as shark, swordfish, grouper, sea trout, and many others (NMFS, 1988). This is an important caveat in comparing differences between imported and U.S. seafood products, because import estimates probably underestimate the consumer's exposure to contaminants as a result of the highly generalized information provided.

The resulting overall yearly per capita seafood consumption estimate, for U.S. and imported products, of about 18 pounds was a little high, but not grossly larger than the official estimate of 15.4 pounds, and well within range to offer the kind of reliability needed for a first-order appraisal of daily seafood intake patterns and exposures. (For proprietary reasons, some NMFS information that would have helped the committee further refine the per capita figures was withheld concerning year-end inventory figures underlying the data in *Fisheries of the United States, 1987*; obtaining a per capita number identical to the NMFS figure would be impossible without additional information).

It can be seen that the leading contributors to U.S. per capita seafood consumption are cod, flounder, tuna, salmon, clams, shrimp, groundfish, and aquatic finfish species from two groupings of miscellaneous fish known as "other marine finfish" and "other freshwater finfish."

Inorganic Contaminant Exposures and Suggested Acceptable Intake

This section employs some of the best available data for estimating seafood-related exposure to inorganic contaminants and compares the indicated national average intakes via commercially marketed seafood with ADIs and related regulatory levels.

This comparison provides a very broad indication of potential risk but is potentially misleading because the variability in exposures to different contaminants, and the variability of contaminant exposures from other nonseafood and noncommercial seafood sources, cannot be estimated from the available data.

Therefore, the committee was unable to arrive at estimates of how many people are likely to exceed the stated ADI intake levels on either an acute or a chronic basis.

Mean levels and ranges of trace metals in the edible portions of certain fish and invertebrates may be obtained from combinations of the data base references previously cited in this chapter.

Three agencies, the FDA, the Australian National Health and Medical Research Council (NHMRC), and the Canadian Health Protection Board (CHPB), have suggested allowable trace-metal levels for seafood and seafood products. The FDA has set an acceptable concentration for mercury based on per capita consumption (FDA, 1982).

The Australian NHMRC has recommended maximum concentrations in seafood for arsenic, cadmium, copper, lead, and zinc (Mackay et al., 1975).

The Canadian government has set action levels based on the contaminant level of edible weight for arsenic, lead, and mercury (DFO, 1989). Alert levels for trace metals in molluscs have been set by the National Shellfish Sanitation Program (NSSP) of the Interstate Shellfish Sanitation Commission.

However, these have no public health significance and are based on mean survey levels in U.S. coastal waters approximately two decades ago.

TABLE : U.S. Landings, Consumption/Live Weight Conversion Factors, and Annual Per Capita Consumption for Different Species

Species	1987 U.S. Landings (10^3 lb)	Conversion Factor (lb consumed/lb round wt)	U.S. Average Annual Per Capita Consumption (lb)[a]
Alewives	20,541		
Anchovies	12,857	1	
Bluefish	15,226	0.43	0.027
Bonito	11,420	0.43	0.020
Butterfish	10,315	0.38	0.004
Cod	229,711	0.5	0.305
Croaker	11,044	0.5	0.023
Cusk	3,064	0.5	0.006
Flounder	199,711	0.45	0.373
Grouper	9,494	0.5	0.002
Hake	56,696	Not given	0.117
Halibut	76,107	0.45	0.116
Herring, sea	207,134[b,c]	0.59	
Jack mackerel	26,671	0.59	0.047
Lingcod	7,027	0.5	0.015
Mackerel	115,103	0.45	0.202
Mullet	30,125	0.5	0.057
Ocean perch	24,184	0.38	0.027
Pollock	597,693	0.5	1.116
Rockfish	117,881	0.5	0.244
Sablefish	102,698	0.5	0.126
Salmon	562,018	0.5	0.510
Scup or porgy	14,295	0.38	0.022
Sea bass	5,083	0.5	0.011
Sea trout	21,871	0.5	0.045
Shark	21,832	0.4	0.036
Snapper	8,908	0.42	0.015

Striped bass	431	0.5	0.001
Swordfish	9,761	0.5	0.016
Tilefish	7,950	0.5	0.016
Tuna	593,444	0.45	1.104
Whiting	34,673	0.45	0.044
Other marine finfish	199,795	0.5	0.376
Other freshwater finfish	117,880	0.5	0.244
Clams	134,357	1.0	0.550
Crabs	386,368	0.183	0.270
Lobsters	51,313	0.22	0.044
Oysters	39,807	1.0	0.164
Scallops	40,773	1	0.165
Shrimp	363,142	0.49	0.657
Squid	82,049[c]	0.6	0.151
Other shellfish	89,702	0.4	0.116
Total	4,655,826		7.362

[a] The numbers in this column are not in all cases equal to the product of the previous two columns because of adjustments for exports and year-to-year inventory changes.
[b] Largely used for fish meal. [c] Largely used for bait.

TABLE: Volume of Imports, Consumption/Live Weight Conversion Factors, and Annual Per Capita Consumption for Different Species

Species	1987 Imports (10^3 lb)	Conversion Factor (lb consumed/lb landed wt)	U.S. Average Annual Per Capita Consumption (lb)
Whole or eviscerated			
Cod, cusk, haddock, and flounder	105,158	0.4875	0.206
Halibut	9,295	0.45	0.017
Salmon	41,902	0.5	0.087
Tuna	787,133	0.45	1.464

Fillets and steaks			
Flounder	73,003	1	0.302
Groundfish	315,418	1	1.303
Other	232,564	1	0.961
Blocks and slabs	403,577	1	1.668
Shrimp	461,173	0.88[a]	1.677
Crabmeat	12,571	1	0.052
American lobster meat	38,974	1	0.129
Spiny lobster meat	41,949	1	0.139
Scallops (meats)	39,934	1	0.165
Analogue products with shellfish	30,539	1	0.126
Other fish and shellfish	98,996	1	0.409
Canned and other products			
Canned herring	5,617	1	0.023
Sardines	65,022	1	0.269
Tuna	211,685	1	0.875
Analogue products without shellfish	4,737	1	0.020
Other	8,797	1	0.036
Abalone	13,974	1	0.058
Crabmeat	7,967	1	0.033
American lobsters	637	1	0.003
Spiny lobsters	136	1	0.001
Oysters	32,668	1	0.135
Shrimp	17,132	1	0.071
Other fish and shellfish	57,579	1	0.238
Pickled or salted			
Cod, haddock, hake, etc.	31,893	1	0.132
Herring	9,991	1	0.041
Other fish and shellfish	16,108	1	0.067
Total	3,176,129		10.707

[a] This is the result of a weighted calculation based on the importation of 310 million pounds of shrimp with the shell on, noting that consumed weight represents 80-85% of landed weight. The rest of the shrimp is assigned a conversion factor of 1.

The United Nations Food and Agricultural Organization in conjunction with the World Health Organization (FAO/WHO, 1972) has suggested provisional tolerable weekly intake limits (PTWIs) for cadmium, mercury, and lead. In addition, there are estimated safe and adequate daily dietary intake (ESADDI) levels for all foods set by the National Research Council (NRC) of the National Academy of Sciences (NRC, 1980).

TABLE: Suggested Trace Metal Action Levels for Seafood and Seafood Products [μg/g (ppm)]

Metal	FDA	ANHMRC[a]	CHPB[b]
Arsenic[c]	-	1.53	3.5
Cadmium	-	2.0	-
Cooper	-	30.0	-
Lead	-	2.0	0.5
Mercury	1.0	-	0.5
Zinc	-	1,000.0	-

[a] Australian National Health and Medical Research Council.[b] Canadian Health Protection Board.[c] As the trioxide.

TABLE: Acceptable Daily Intake Limits of Trace Metals as Suggested by FAO/WHO (PTDI) and NRC-NAS (ESADDI) Compared with Mean U.S. Daily Intake of Trace Metals Reported by FDA Total Diet Studies (μg/day)

Metal	FAO/WHO PTDI[a]	NRC-NAS ESADDI	FDA[b] Mean Daily Intake	Mean Daily Intake from 15-lb/yr U.S. Catch[c]
Antimony[d]			15	
Arsenic	182[e]	-	45	82 (43.6)[f]
Cadmium	57-72	-	15	2.0
Chromium	-[e]	50-200	-	3.7
Copper	-	2,000-3,000	-	51
Lead	429	-	41	10[i]
Mercury[g]	33-43	-	3.9	2.1
Nickel				5.6
Selenium	-	50-200	152[h]	14
Zinc	15,000	-	17,700[h]	

[a] Calculated from the PTWI; 70-kg human.
[b] Total Diet Studies for 25-30-year-old males (FDA market basket program); Gunderson et al. (1988) except as otherwise noted.
[c] Calculated from data in Hall et al. (1978); for the breakdown of average levels for species, see Tables 6-10-6-18, which detail the distributions of species-average concentrations.
[d] EPA RfD = 4×10^{-4} mg/kg/day = 28 μg/day.
[e] Arsenic trioxide; not agreed upon but estimated.
[f] Data in parentheses are FDA seafood measurements given in Gunderson et al. (1988).
[g] Lowest value is methylmercury, highest total mercury.
[h] From the older total diet data of Gartrell et al. (1985).
[i] A study coauthor believes this result may be too high because of contamination.

Finally, although most of the hazards from these inorganic elements are chronic in nature, one can get a feel for potential short-term dosage from shellfish by multiplying the ranges of contaminant levels found in different surveys by an assumed 250-g intake at a single sitting. These findings are discussed below in the separate subsections for each element. The data base used for this was Cappuzo et al. (1987) for the northeastern coastal region (review 1972-1983, 5 species); Landolt et al. (1985) for Puget Sound (review 1982-1985, 4 species, n = ?); Landolt et al. (1987) for Puget Sound (1985-1986, 11 species, $n = 64$); Murphy (1988a-c) for Chesapeake Bay (bluefish, $n = 71$; striper, $n = 35$). Chromium is mentioned only because oral intake of amounts as small as 50 μg of potassium dichromate may exacerbate existing chromium hypersensitivity, and reported levels in finfish and squid would fulfill that criterion (Haines and Nieboer, 1988) if they were in the hexavalent form (which is probably not likely).

Use of NMFS Survey of Trace-Element Data

Available data from the NMFS survey of trace elements (Hall et al., 1978) are now examined in relation to these advisory levels.

Antimony

The committee was surprised to find that the average U.S. consumption of antimony from seafood indicated by the Hall et al. (1978) data base is within a factor of two of EPA's reference dose (RfD) value intended to protect against systemic toxicity (EPA, 1986b). It is unclear what forms of antimony are present in seafood or whether those forms have toxicity similar to the forms that gave rise to the RfD. Antimony is below arsenic on the periodic table, and antimony dust appears to be carcinogenic for the lung when tested by inhalation (Groth et al., 1986). The committee is not aware of any long-term testing by oral routes. According to the abstract of an EPA document, increases in spontaneous abortions and premature births have been reported for pregnant female workers exposed to antimony (EPA, 1981). Further investigation of antimony's potential for hazard via seafood seems warranted.

Arsenic

Clearly, seafood is an important source of overall arsenic exposure in the diet, although as mentioned earlier, the organic forms of arsenic that appear to predominate in the seafood species studied in detail are considerably less hazardous than inorganic forms of arsenic.

TABLE: Distribution of Species-Average Antimony Concentrations, Weighted by the Percentage Contribution of Each Species to 1970 Human Consumption

Range (ppm) for Species Average Consumption	Midpoint (ppm) of Range	Percentage of U.S. Catch Intended for Human Consumption
Finfish, muscle		
0.5-0.6	0.55	3.6
0.6-0.7	0.65	17.3
0.7-0.8	0.75	14.9
0.8-0.9	0.85	26
0.9-1.0	0.95	1.7
1.0-2.0	1.5	0.5
All finfish	0.764	64
Molluscs, edible tissues		
<0.1	0.05	3.4
0.5-0.6	0.55	0.4
0.6-0.7	0.65	0.6
0.7-0.8	0.75	0.1
0.8-0.9	0.85	1.9
0.9-1.0	0.95	1.1
1.0-2.0	1.5	0
All, average	0.469	7.5
Crustaceans, edible tissues		
0.2-0.3	0.25	0.1
0.8-0.9	0.85	8.7
0.9-1.0	0.95	11.7
1.0-2.0	1.50	7.8
All crustaceans	1.07	28.3
Grand weighted average (ppm)	0.83	99.8[a]
Daily dose for 15 lb/ yr (μ/ day)	15.4	

[a] Total is not 100% because of rounding error and because some species were excluded from the study.

SOURCE: Hall et al. (1978).

TABLE: Distribution of Species-Average Arsenic Concentrations, Weighted by the Percentage Contribution of Each Species to 1970 Human Consumption

Range (ppm) for Species Average Consumption	Midpoint (ppm) of Range	Percentage of U.S. Catch Intended for Human Consumption
Finfish, muscle		
0.6-0.7	0.65	0.2
1.0-2.0	1.5	2.6
2.0-3.0	2.5	35
3.0-4.0	3.5	13.8
4.0-5.0	4.5	3
5.0-6.0	5.5	2.8
6.0-7.0	6.5	0.3
7.0-8.0	7.5	5
8.0-9.0	8.5	0.1
9.0-10.0	9.5	0.2
10.0-20.0	15	0.7
20.0-30.0	25	0.2
All finfish	3.54	63.9
Molluscs, edible tissues		
2.0-3.0	2.5	4
3.0-4.0	3.5	3.2
4.0-5.0	4.5	0.1
10.0-20.0	15	0.1
All, average	3.13	7.4
Crustaceans, edible tissues		
3.0-4.0	3.5	1.8
4.0-5.0	4.5	14.2
5.0-6.0	5.5	2.5
6.0-7.0	6.5	5.8
9.0-10.0	9.5	2
10.0-20.0	15.0	0.5
20.0-30.0	25.0	1.5
All crustaceans	6.56	28.3
Grand weighted average (ppm)	4.37	99.6
Daily dose for 15 lb/ yr (µg/ day)	81.5	
SOURCE: Hall et al. (1978).		

Preliminary examination of muscle levels in fish from the areas of interest from other data bases indicates a range of 0.25-20.7 ppm in finfish and up to 15.9 ppm in squid. The estimate of intake in micrograms for a 250-g serving ranges from 62.5 to 5,157. In contrasting this with the estimated FAO PTDI of 182 μg (calculated as the trioxide to be 2 μg/kg/day for a 69.1-kg human), the highest possible acute daily value exceeds the PTDI by 4,975 μg, roughly a factor of 25. No ESADDI is given by the NRC.

TABLE: Distribution of Species-Average Cadmium Concentrations, Weighted by the Percentage Contribution of Each Species to 1970 Human Consumption

Range (ppm) for Species Average Consumption	Midpoint (ppm) of Range	Percentage of U.S. Catch Intended for Human Consumption
Finfish, muscle		
<0.1	0.05	63.8
0.1-0.2	0.15	0.1
All finfish	0.050	63.9
Molluscs, edible tissues		
0.1-0.2	0.15	3
0.2-0.3	0.25	1.2
0.7-0.8	0.75	0.6
0.9-1.0	0.95	2
1.0-2.0	1.5	0.6
2.0-3.0	2.5	0.1
All, average	0.567	7.5
Crustaceans, edible tissues		
<0.1	0.05	9.8
0.1-0.2	0.15	17
0.2-0.3	0.25	1.5
0.3-0.4	0.35	0.1
All crustaceans	0.12	28.4
Grand weighted average (ppm)	0.11	99.8
Daily dose for 15 lb/yr (μg/day)	2.0	
SOURCE: Hall et al. (1978).		

Cadmium

The overall seafood-related consumption of cadmium estimated as 2 μg/day is a modest, but not completely insignificant (6%), portion of the overall dietary exposure estimated from the total diet program. Within seafood, cadmium is relatively highly concentrated in molluscan shellfish. Muscle levels in fish and squid from the designated data base range from 0.0006 to 0.63 ppm. For a 250-g serving, intake would vary from 0.15 to 157.5 µg. The PTDI for cadmium is 57-72 µg, approximately one-half to one-third the highest calculated intake. No estimated safe daily dietary intake is suggested by the NRC.

TABLE: Distribution of Species-Average Chromium Concentrations, Weighted by the Percentage Contribution of Each Species to 1970 Human Consumption

Range (ppm) for Species Average Consumption	Midpoint (ppm) of Range	Percentage of U.S. Catch Intended for Human Consumption
Finfish, muscle		
<0.1	0.05	0.3
0.1-0.2	0.15	51
0.2-0.3	0.250	12.5
All finish	0.169	63.8
Molluscs, edible tissues		
0.1-0.2	0.15	0.6
0.2-0.3	0.25	0.7
0.3-0.4	0.35	6.1
All, average	0.324	7.4
Crustaceans, edible tissues		
0.1-0.2	0.15	5.4
0.2-0.3	0.25	22.9
All crustaceans	0.23	28.3
Grand weighted average (ppm)	0.20	99.5
Daily dose for 15 lb/yr (µg/day)	3.7	
SOURCE: Hall et al. (1978).		

Chromium

The overall consumption of chromium from seafood indicated by the Hall et al. (1978) data appears well below the applicable ADI. Chromium levels in the muscles of finfish and squid for the areas of interest varied from 0.35 to 2.0 ppm. Calculated short-term intake levels are therefore 87.5-5,000 µg for a 250-g portion. The FAO has no suggested PTWI. There are no action levels reported for chromium in seafood. However, the ESADDI for chromium is 50-200 µg, which is far lower than the highest calculated intake (by a factor of 25). Again, whether a rare acute exposure of this magnitude is problematic is unclear.

Lead

In the light of recent research indicating developmental toxicity even from levels of lead considered to be in the normal range, this would be worthy of further study. However, in this case, the committee has been cautioned by one of the coauthors of the Hall study (M. Meaburn) that the analyses were conducted under

TABLE: Distribution of Species-Average Lead Concentrations, Weighted by the Percentage Contribution of Each Species to 1970 Human Consumption

Range (ppm) for Species Average Consumption	Midpoint (ppm) of Range	Percentage of U.S. Catch Intended for Human Consumption
Finfish, muscle		
0.3-0.4	0.35	9.9
0.4-0.5	0.45	33.8
0.5-0.6	0.55	9.4
0.6-0.7	0.65	10.7
All finfish	0.48	63.8
Molluscs, edible tissues		
0.4-0.5	0.45	0.4
0.5-06	0.55	3.4
0.6-0.7	0.65	3
0.7-0.8	0.75	0.7
0.8-0.9	0.85	0.1
All, average	0.61	7.6
Crustaceans, edible tissues		
0.4-0.5	0.45	0
0.5-0.6	0.55	3.5
0.6-0.7	0.65	18.5
0.7-0.8	0.75	6.2
0.8-0.9	0.85	0.1
All crustacea	0.66	28.3
Grand weighted average (ppm)	0.54	99.7
Daily dose for 15 lb/yr (µday)	10.1	

NOTE: One of the coauthors of the Hall et al. (1978) study considers these analyses overstated because of contamination problems.
SOURCE: Hall et al. (1978).

TABLE: Distribution of Species-Average Mercury Concentrations, Weighted by the Percentage Contribution of Each Species to 1970 Human Consumption

Range (ppm) for Species Average Consumption	Midpoint (ppm) of Range	Percentage of U.S. Catch Intended for Human Consumption
Finfish, muscle		
<0.1	0.05	30.7
0.1-0.2	0.15	26.1
0.2-0.3	0.25	3.9
0.3-0.4	0.35	2.2
0.4-0.5	0.45	0
0.5-0.6	0.55	0.7
0.6-0.7	0.65	0
0.7-0.8	0.75	0.1
All finfish (except swordfish)[a]	0.120	63.7
Molluscs, edible tissues		
<0.1	0.05	7.5
Crustaceans, edible tissues		
<0.1	0.05	13.7
0.1-0.2	0.15	13.2
0.2-0.3	0.25	1.4
All crustacea	0.107	28.3
Grand weighted average without swordfish (ppm)	0.111	99.5
Swordfish	0.95[b]	0.107[c]
Grand weighted average with swordfish (ppm)	0.112	
Daily dose for 15 lb/yr (µg/day)	2.09	

[a] According to one of the coauthors, for "policy reasons" swordfish was not included in the species sampled for the Hall et al. (1978) survey.

[b] Average from FDA surveillance samples.

[c] Based on total swordfish consumption of about 4 million lb/year (of about 9.7 million lb round weight landed). This represents an average of about 0.016 lb per person per year. [Landings data from NMFS (1988); consumed weight/round weight ratio taken from other NMFS data.]

SOURCE: Hall et al. (1978).

Muscle levels in finfish and squid from the designated data base range from 0.008 to 12 ppm. Calculated intake for a 250-g serving would be 2.0-3,000 μg. The higher range is certainly skewed by fish taken from Boston Harbor. A more realistic range would be from 0.008 to 2.3 ppm, resulting in an intake of 2.0-575 μg. The PTDI for lead is 429 μg. In the first instance, the ADI is exceeded by a factor of 6; in the second, by 1.3. No ESADDI is given. The normal estimated daily lead intake for children is less than 0.3 μg (Green et al., 1978).

Mercury

The estimated population average consumption of mercury from aquatic animals of about 2.1 μg/day represents over half of the mercury estimated to be in the diet as a whole for the age and sex group (25-50-year-old males) with the highest daily mercury consumption in the FDA Total Diet Study. Aquatic organisms probably make a larger aggregate contribution to the methylated portion of total mercury (total mercury, of course, includes inorganic mercury, which has very different risks) (Gunderson, 1988).

The 2.1-μg/day average is more than an order of magnitude below the official ADI of 33 μg/day (as methylmercury). As indicated earlier in this chapter, in one of its specific studies of FDA regulations the committee attempted to assess the degree of actual protection offered by this spread (and the degree of "safety" incorporated into the ADI itself) by analysis of data on the population distributions of internal dosage (as measured in blood or hair) and the interindividual variation in susceptibility to different effects in adults and developing fetuses.

Swordfish can routinely achieve even higher concentrations than the official 1-ppm guideline (which reportedly is not enforced in Massachusetts and perhaps elsewhere). The suggested PTDI for total mercury is 43 μg, which is exceeded by a factor of 6 in a serving containing 1 ppm. The FDA allowable level is 1.0 ppm and the Canadian action level is 0.5 ppm. The NRC has not suggested an estimated safe daily dietary intake level.

TABLE: Distribution of Species-Average Selenium Concentrations, Weighted by the Percentage Contribution of Each Species to 1970 Human Consumption

Range (ppm) for Species Average Consumption	Midpoint (ppm) of Range	Percentage of U.S. Catch Intended for Human Consumption
Finfish, muscle		
0.1-0.2	0.15	0.5
0.3-0.4	0.35	2.3
0.4-0.5	0.45	21.2
0.5-0.6	0.55	8.3
0.6-0.7	0.65	4.9
0.7-0.8	0.75	2.1
0.8-0.9	0.85	5.4
0.9-1.0	0.95	1.5
1.0-2.0	1.50	17.7
All finfish	0.819	63.9
Molluscs, edible tissues		
0.3-0.4	0.35	3.2
0.4-0.5	0.45	0.8
0.5-0.6	0.55	0.7
0.7-0.8	0.75	2.3
0.8-0.9	0.85	0.5
All, average	0.54	7.5
Crustaceans, edible tissues		
0.2-0.3	0.25	0.4
0.3-0.4	0.35	5.1
0.4-0.5	0.45	2.1
0.5-0.6	0.55	1.1
0.6-0.7	0.65	11.9
0.7-0.8	0.75	6.2
0.8-0.9	0.85	1.4
1.0-2.0	1.50	0.1
All crustacea	0.61	28.3
Grand weighted average (ppm)	0.737	99.7
Daily dose for 15 lb/yr (µg/day)	13.75	

SOURCE: Hall et al. (1978).

Selenium

The committee's overall average calculated selenium daily intake of slightly less than 14 µg/day represents somewhat less than 10% of the estimate of total dietary loading, and between 7 and 30% of NRC's ESADDI.

Selenium levels in teleosts (fish with bones) seldom exceed 1 ppm (Sorensen et al., 1984). The range of selenium levels from this data base was 0.0-0.49 ppm and came from determinations in Puget Sound animals only. Selenium is very much a regional, freshwater problem (most notably in a wildlife refuge that receives extensive irrigation tile drainage in California), and the chosen data base for this preliminary report does not reflect the apparent regional risk. Calculated intake levels are therefore somewhat lower than found in problem areas and range from 0.0 to 122.5 µg.

This is less than the high range of the NRC ESADDI of 50-200 ¼g. When the data base includes site-specific studies, values for skeletal muscle may vary from 0.5 to 12.9 ppm (Cumbie and Van Horn, 1978; Sorensen et al., 1984). Calculated values of intake using this data base are 125-3,225 µg, the highest value exceeding the PTDI by a factor of 15. The Fish and Wildlife Service (F&WS) National Pesticide Monitoring Program (1980-1981) also reports higher values for selenium, which range from 0.09 to 2.47 ppm; however, these values are for the whole fish and not specifically for skeletal muscle. Thus, their use in assessing risk is devalued. The FAO/WHO Joint Committee has not suggested a PTWI for selenium. The toxic chronic dose of selenium for man is estimated as 2,400 to 3,000 µg/day (Wilber, 1983).

Estimates Of Cancer Risks From Organic Contaminants

Once the committee established per capita consumption patterns for a variety of seafood species as outlined earlier, average concentrations of contaminants calculated from the available FDA surveillance and compliance data were used to estimate the average amounts of industrial organic chemicals, pesticides, and selected inorganics delivered to U.S. consumers via commercially marketed seafood.

As discussed earlier, these FDA surveillance and compliance data have quite a few weaknesses. Improvements in monitoring can be made by the responsible agencies. Yet, despite these weaknesses, the committee believes that the data provide a realistic picture of national aggregate organic chemical exposure from commercially marketed seafood. These are the only data that enable even a preliminary estimate of national aggregate cancer risks from organic carcinogens.

Among the weaknesses in the FDA data are the following:

The detection limits used by FDA regulatory laboratories result in a generally gross underestimate of actual contaminant concentrations in seafood. For example, fish shown by other studies to be contaminated with low-level PCBs appeared to have no contamination in the FDA data set. This is true of California species such as bonito, mackerel, squid, and white croaker, which probably do have consistently low levels of PCBs according to studies performed by the Southern California Coastal Water Research Organization (Gossett et al., 1982). The committee believes that because of their frequency in the aquatic food chain and in some areas of the seafood supply, PCBs constitute an important contaminant even at levels below the present regulatory detection limits. Many contaminants probably go undetected by the FDA regulatory monitoring program because the laboratory detection limits used for enforcement of regulations have levels 5-10 times higher than the limits used in the Total Diet Study analyses (Gunderson, 1988, pp. 1200-1209).

There are significant problems of misidentifying some fish species. The FDA lists the Great Lakes chub under whitefish; yet according to the 1988 *Fish List* (FDA, 1988), chub and whitefish are not the same species. This has added a great deal of confusion to the FDA data base.

There is a decided lack of precise identification of species. In the FDA data base, there is only one classification of "mullet." In fact, however, 11 different species of mullet may be sold for food in interstate commerce. It would be helpful to know the precise species being sampled by FDA. In that way, it could be determined whether certain species of mullet have greater contamination than others. Another example is FDA's sampling of "croaker." There are more than 20 species of croaker that are sold for food fish according to the 1988 *Fish List* (FDA, 1988); yet it is unclear which croakers are actually being sampled and from which of the nation's coastlines. Because contaminant levels among different croaker species may vary dramatically, the committee believes that such information would prove extremely helpful in refining the understanding of which fish species, in particular, contribute the highest dietary exposure to toxic chemicals.

There is also a tremendous need for precise geographic information. The FDA should determine where each sample has been caught.

Because of the limited number of samples, FDA data are clearly inadequate and present absolutely no meaningful overall information on contamination of many species. Some aquatic organisms are sampled sporadically.

Many aquatic species are never analyzed for mercury concentrations.

Many finfish are never analyzed for inorganic contaminants, such as arsenic, cadmium, lead, and selenium, in edible portions.

Very few shellfish are analyzed for organic chemical contamination or for other contaminants of concern such as organotins and polycyclic aromatic hydrocarbons (PAHs) even though it is known that bivalves may well be contaminated with such chemicals.

It is uncertain for which pollutants FDA routinely tests, and at what laboratory detection limits. For example, the FDA Total Diet Study was able to detect pesticides such as chlorpropham, dacthal, diazinon and malathion on 45 occasions when analyzing 48 composite samples of seafood dishes such as cod and handdock fillets, canned tuna, and shrimp–pesticides that the FDA industrial chemical and pesticide monitoring program rarely, if ever, detected among these seafood items in some five years of sampling (Gunderson, 1988). This may occur because these pesticides are applied to the food product after harvest and during processing, because the analytical methods in use are not designed to detect these particular chemicals, or because the laboratory detection limits in the Total Diet Study have quantitation levels that are 5-10 times lower than the limits used in FDA enforcement of regulatory limits (Gunderson, 1988). In any event, FDA regulatory data appear to underestimate consumers' exposure to a variety of organic and inorganic contaminants including pesticides, metals, and PAHs.

The disparity in sampling detection limits between the Total Diet Study and regular FDA seafood monitoring programs raises serious doubts as to whether the FDA sampling program is accurately estimating the actual concentrations of industrial chemicals and pesticides in the seafood supply. Indeed, the FDA laboratories' relatively high detection levels–particularly for PCBs and dieldrin, which can occur in seafood at levels below the current detection limits–lead to so-called nondetected zero values when the PCB concentration might actually be several parts per billion. Because of the relatively high carcinogenic potency of dieldrin and PCBs, and their frequency in the seafood supply, findings of several parts per billion in widely consumed aquatic species might add appreciably to overall exposure and risk.

Nevertheless, FDA data can be used to provide a rough estimate of daily dietary exposure and at least a highly tentative indication of potential risk.

Using such calculations to determine the nationwide daily per capita intake of seafood species may well tend to obscure individual risks associated with industrial chemicals. Certain consumer subgroups probably favor certain seafood over others, whether their reasons be cultural, taste, health, or simply economic. Thus, any one individual or group of individuals may have a handful of favorite seafood items that are eaten repeatedly, which means that exposure is not spread evenly throughout the population but tends to be concentrated among certain subgroups. Based on landings data, a significant amount of freshwater fish, classified under the NMFS data as "other freshwater finfish," is being eaten; consumers of freshwater finfish probably are not spread out among the approximately

242 million citizens. Also, this category contains many of the most contaminated finfish.

It can be seen that the overall estimated cancer risk is dominated by the (highly uncertain) estimate for PCBs, dieldrin, DDT, and dioxin. Nevertheless, the overall risk as assessed here is clearly not negligible, if these cancer potency estimates are at all close to the mark, and there is no way to know at present that they are not.

For arsenic and cadmium, the estimates from the FDA surveillance data are lower than the Hall et al. (1978) data set but are generally within twofold. This is probably because of the lack of monitoring data for these inorganic contaminants in the edible portions of finfish. For chromium, load, and especially selenium, there are larger differences, with the FDA data set in all cases providing the lower estimates. It is quite possible that the paucity of coverage of some important species in the FDA data set, the probably higher detection limits, and the committee's practice of assigning zero contaminant values to both samples below detection limits and species that were not sampled for a particular contaminant, lead to a downward biasing of the exposure estimates from the FDA data. It would therefore not be surprising to find that some of the estimates of organic contaminants in Table may have a similar downward bias.

Based on the FDA data set, the six U.S. commercially caught seafood products that appear to present the greatest daily per capita aggregate PCB exposures and their upper-confidence-limit cancer risks are other freshwater finfish (5.4 $\times 10^{-6}$), bluefish (6.3×10^{-7}), mackerel (2.5×10^{-7}), and sea trout, mullet, and scup or porgy (each 1.1×10^{-7}). The reader should be cautioned that these data are not adequate to fully characterize PCB levels for individual species in a nationally representative way. Nevertheless, the result that probably does have some significance is that the bulk of the PCB risk is concentrated in a relatively obscure category of "other freshwater fish." This group includes fishery products from the Great Lakes and other inland waterways. To the extent that PCBs are considered to be a problem, the bulk of contamination comes from a minor, identifiable fraction of the overall seafood in commerce.

The U.S. commercially caught seafood species presenting the greatest daily per capita dieldrin exposures and upper-confidence-limit cancer risks to the general population are other freshwater fish (4.7×10^{-7}), and mullet (1.0×10^{-8}).

Comparison of Imported and Domestic Seafood

This analysis is also based on FDA monitoring data. Information on individual imported fish is statistically insignificant, but when taken as a whole the FDA monitoring program appears to suggest some relevant differences in terms of contamination of the domestic supply versus that of the imported supply. Imported seafood, overall, appears less contaminated than domestically landed seafood. The committee believes that this overall difference is largely the result of contamination of U.S. freshwater finfish products. If freshwater finfish were not part of the U.S. seafood supply, U.S. saltwater fish probably would not differ much from imported fish. This could be established with greater precision by a more directed and well-designed sampling effort. Such an effort would be valuable in protecting the consumer and more cost effective than a visual inspection program. In fact, the hypothesis that significant differences in contamination exist between species of fish, as well as between domestically landed and imported seafood, offers a rationale to help consumers reduce toxic exposures by accurately identifying species and the geographic origin of seafood products offered for sale. This could aid consumers in choosing the least contaminated products.

Overall, in only a few cases of contaminated finfish, did imports provide a greater exposure for the consumer than the domestic catch. The committee realizes, however, that this observation may be due to the fact that import consumption patterns were not narrowed down to individual species, that many imported species were not always identified, and that many were not analyzed.

Those cases in which imports had higher contaminant levels included chemicals such as lindane, tecnazene, cadmium, and mercury. As for cadmium, canned tuna accounts for much of the consumer's exposure. Other, lesser contributors of cadmium include imported scallops. For mercury, the biggest contributors are shark, swordfish, and tuna. An estimation of the total mercury exposure from these imports could not be determined because NMFS import data do not detail specific consumption figures for imported shark and swordfish.

Among imports, salmon and herring accounted for primary sources of PCBs in the diet. Yet, only four samples of herring were taken by the FDA monitoring program to be sampled for organic contamination between 1983 and 1987. This lack of data for individual species is disconcerting, especially when there are definite warning signals indicating that more data gathering would be warranted. The PCB concentration in imported sea herring ranged from nondetectable to 0.36 ppm with three positive detections out of the four samples – a significant positive percentage and range. Furthermore, 11 different industrial pollutants and pesticides were found in these four imported sea herring samples, and although the levels were below FDA action guidelines, these four samples with 27 individual industrial chemicals and pesticide residues warrant further sampling.

Approximately 12 samples of imported salmon were analyzed for organic contaminants between 1986 and 1988. Two of the nine samples from Norway were positive for PCBs, tecnazene, and lindane. Because so little is known about noncancer long-term health consequences of PCBs and other toxic exposures, it seems prudent to reduce PCB exposure in any way possible, and with farm-raised products, the reduction of PCB and other organic chemical contamination should be possible.

Import data are virtually comprised entirely of finfish samples. Very few samples of the imported seafood examined consisted of shellfish. There simply is not enough information available for the committee or anyone to determine the extent of contamination of imported shellfish.

Based on the data available, the biggest single potential risk to consumers in seafood is posed by exposure to PCBs. Yet, whereas the upper-confidence-limit risk per capita from PCBs alone (not counting DDT, dieldrin, and dioxin) is 5.8×10^{-5} for domestically landed finfish, the risk from imports appears significantly lower, slightly less than 3×10^{-6}. For the next highest cancer risk chemical, dieldrin, the U.S. catch typically provides a risk to the average consumer of approximately 1×10^{-5}. The risk per capita from dieldrin in imported fish also appears much smaller, less than 1×10^{-6}. These findings should not be overgeneralized nor, based on present knowledge, can one assume that imported seafood is safer than domestically landed seafood.

Exposures from Sport, Subsistence, and Tribal Fishing

Noncommercial fishing is a significant source of overall fish for human consumption in the United States – estimated at 3-4 pounds per person on the average. It is even more significant as a vehicle for the delivery of bioaccumulating chemicals for the following reasons:

1. Noncommercial seafood harvesters often concentrate their fish/shellfish harvesting within specific areas, some of which may be heavily contaminated.
2. A minority of recreational and subsistence seafood harvesters engage in this activity very frequently and consume relatively large amounts of fish (Humphrey, 1983a,b).
3. Some of the most significant recreational/subsistence caught species also happen to be among those with the largest concentrations of contaminants (e.g., northeastern bluefish; southern California white croaker; Great Lakes trout, walleye, and salmon).
4. In some states (e.g., New York), relatively uncontaminated areas appear to be difficult to find (New York State DEC, 1987).

The primary mechanism by which state governments have alerted consumers about the potential risk of contaminated finfish is the fish consumption advisory (Zeitlin, 1989). In fact, among the 30 U.S. coastal and Great Lakes states, Hawaii, and

Alaska (with the exception of Georgia, which did not respond to the survey), 2,094 advisories were issued for coastal marine water, estuaries, rivers, and inland waters. Overall, the vast majority of the advisories (87%) were issued for freshwater fish. Among the states indicated, Alaska, Alabama, New Hampshire, and Oregon each issued one advisory between 1984 and 1987. On the other hand, Minnesota issued 665. As Zeitlin (1989) notes Consistently high numbers of advisories were issued by the Great Lakes states. Even if advisories are posted to warn anglers, some persons may not heed these warnings because they do not get sick immediately from the consumption of contaminated seafood. There is no way of knowing how effective they are. The effectiveness of advisories in convincing the general public to voluntarily alter their preparation and consumption habits is, for the most part, unknown. Very few epidemiological or social studies have been conducted. The inherent difficulty in changing pleasurable human behavior is not a novel concept and public perception of risk is often difficult to measure. Adherence to hazard warnings often depends on how much confidence the public has in the credibility of the agency issuing the warnings.

Thus, anglers may be constantly re-exposed to contaminants. Further, many of the nation's most contaminated areas are relatively accessible to the recreational seafood gatherer (e.g., estuarine regions and freshwater bodies). In addition, many "recreational" anglers may actually be subsistence anglers who fish as an important means of supplementing their diet, and they may share their catch with family members and friends.

According to Zeitlin, methods used to communicate potential risks include media announcements, printed brochures, posting notices in public places, and information in fishing license applications. Members of the medical community were contacted to disseminate information to patients in New York, Maine, Michigan, Minnesota, and Wisconsin. Indiana, New York, New Jersey, Michigan, Minnesota, Washington, and South Carolina have organized public outreach and education programs. Finally, contamination of the nation's recreational fishery appears to be pervasive, and although extensively documented as well as possible to date, the contamination is probably not very thoroughly documented – particularly for freshwater fishing.

Examination indicated that PCBs accounted for 43% of all advisories issued. Mercury accounted for 40%, followed by chlordane and dioxin at 8% and 2%, respectively. The remaining 7% was issued for chemicals such as dieldrin, kepone, DDT, PAHs, heptachlor epoxide, petroleum compounds, selenium, and chlorinated benzenes (Zeitlin, 1989). Advisories issued according to families of freshwater fish were Percidae, including yellow perch and walleye, 21%; Salmonidae, 16% (trout, 11%; salmon, 5%); Centrachidae (sunfish and bass), 13%; Esocidae (pike), 12%; and Cyprinidae (carp), 11%. The remaining advisories were divided among Ictularidae (bullhead catfish), 10%; Catostomiae (suckers), 5%; Anguillidae (freshwater eels), 4%; striped bass and white perch, 4%; and other species, 4%. Based on regional case studies, recreational anglers probably eat far more seafood than nonrecreational seafood consumers. In the Puget Sound study, average daily consumption was 12.3 g (Landolt et al., 1987). The highest consumption rate for a smaller number of fishermen and women was estimated to be 95.1 g/day. These amounts are somewhat similar to findings for Los Angeles area recreational anglers.

Puget Sound Contaminated Recreational Fish Study

During 1986-1987 a broad-scale survey was sponsored by EPA Region X and supported by the Washington State Department of Ecology and the Department of Social and Health Services (DSHS) to characterize potential human health risks associated with chemical contaminants in Puget Sound seafood (Landolt et al., 1987). Research focused entirely on recreational harvesters who frequently collect fish, shellfish, or edible seaweed for use in personal consumption. Data were available for chemical contaminants in fish, shellfish, and macroalgae from 22 locations in Puget Sound. Fish, shellfish, and macroalgae evaluated in the risk assessments of chemically contaminated seafood included pelagic species such as coho salmon, chinook salmon, Pacific hake, striped perch, rockfish, and sablefish; bottom-feeding species such as Pacific cod, walleye, pollock, and tomcod; bottom fish such as English sole, starry flounder,

Pacific sanddab, rock sole, flathead sole, and buffalo sculpin; shellfish such as Dungeness crab and rock crab; bivalves including heart cockles, bent nose clams, sand clams, soft-shell clams, littleneck clams, butter clams, Manila clams, and horseneck clams; also examined were market squid, brown algae (kelp), green algae (also known as sea lettuce), and red algae (also known as nori). Thus, the study was both very thorough and very complete. The chemicals of highest concern included carcinogens such as arsenic, PAHs including benz[*a*]anthracene, benzo[*a*]pyrene, benzo[*b*]fluoranthene, and chrysene; PCBs, alpha-hexachlorocyclohexane (HCH); and DDTs. Noncarcinogenic chemicals of concern included cadmium, lead, and mercury. Interestingly, FDA tolerances or action levels existed for only three of the eight chemicals of highest concern: mercury, PCBs, and DDTs. Overall, the average concentrations of these substances in the various species and categories of seafood were lower than the FDA tolerances or action levels for all locations in Puget Sound, with the notable exception of the concentration of PCBs (2.06 ppm) in English sole from Elliott Bay.

The average consumption rate for Puget Sound recreational anglers was 12.3 g daily. The highest consumption rate for a smaller number of anglers was estimated to be 95.1 g of fish per day from Puget Sound. In this study, upper-confidence-limit cancer risks for cumulative exposure to arsenic and the four organic chemicals of concern ranged from 2×10^{-4} for average anglers to 4×10^{-3} for those most highly exposed. The relatively high, assessed upper-confidence-limit cancer risk derived primarily from potential exposure to PCBs in fish from specific locations in Puget Sound such as Commencement Bay, Elliott Bay, Manchester and Sinclair Inlet. If the contribution of PCBs is excluded, the remaining risk is less than 10^{-5}. No effort was made in this study to quantitatively evaluate the potential for noncancer risks.

It is also quite apparent from this study that recreational anglers have favorite spots and that this is important information for the distribution of risks among the community. As Landolt et al. (1987) note, "the potential cumulative health risks associated with the consumption of chemically contaminated seafood may exceed [one in 10,000 cancers in excess] for certain locations in Puget Sound."

Regional Studies of Contaminants

Regional studies conducted in New York, Massachusetts, the District of Columbia, Alabama, Kansas, California, and Wisconsin indicate levels of inorganic and organic chemicals in recreationally important fish species (EPA, 1983, 1987; Kansas, 1988; New York State DEC, 1987; Rosen, 1989). Using data generated by these regional studies, one can tentatively estimate upper-confidence-limit cancer risks that might be attributable to dioxin contamination of fish in selected areas. Given an average-size serving (based on recommendations of the Tolerance Assessment System in the EPA Office of Pesticide Programs) of approximately 114 g (0.25 pound) (PTI, 1987), if one were to eat a one-quarter pound serving of large-mouth bass with an average of 11.3 parts per trillion dioxin once a month for a lifetime, the upper-confidence-limit cancer risk implied by the EPA cancer potency factor would be 1×10^{-4}. Therefore, cases may exist in which contamination with dioxin could be a potentially serious hazard for recreational or subsistence anglers.

As suggested earlier, another potential problem residue is methylmercury. Michigan's 1989 fish consumption advisory for anglers suggested that consumption of certain fish from inland lakes statewide be restricted because of mercury found in fish from Upper and Lower Peninsula lakes; approximately 60 of Michigan's inland lakes were tested since 1983, which indicated that about three out of four had at least some fish with mercury exceeding the Michigan Department of Public Health (MDPH) level of concern for state citizens (Michigan, 1988). Pregnant women, nursing mothers, women who intend to have children, and children age 15 or under were advised to eat no more than one meal per month of rock bass, crappie, and yellow perch over 9 inches in length. They were advised not to eat any large-mouth bass, small-mouth bass, walleye, northern pike, or muskie from inland lakes. Others were advised to eat no more than one meal per week of these fish.

3

Nature of Public Health Risks

The principal source of data on the incidence of seafood-borne illness in the United States is the Centers for Disease Control (CDC). Its compilations are accumulated passively from state reports of food-borne outbreaks. An outbreak is an incident involving two or more sick individuals, except for botulism and certain chemical poisonings in which one sick individual constitutes an outbreak. A case is a single ill person. Unfortunately, not all states report each of the major types of seafood-borne disease, and within states there is considerable underreporting of incidents for a variety of reasons. Thus, CDC data may not be representative of actual disease occurrence and may omit altogether important seafood-originated disease outside the reporting format. The CDC information is useful when supplemented by other data on the occurrence of pathogens, an understanding of the patterns of seafood harvest and processing, a knowledge of the mechanisms of disease development, and – where available – independent epidemiological data. The CDC data cannot be used to estimate risks from chemical contamination, because no disease outbreaks have yet been reported from this cause.

Seafood-borne illness reported by CDC in the 10-year period 1978-1987 totaled 558 outbreaks involving 5,980 cases. However, fish and shellfish constitute only 10.5% of all outbreaks and 3.6% of all cases when food-borne illnesses from all foods are considered. The number of people made ill from beef (4%) and turkey (3.7%) exceeds the seafood total, whereas pork (2.7%) and chicken (2.6%) are slightly lower. If shellfish (2.3%) and fish (1.2%) are considered separately, the number of reported cases from each is lower than for any animal meat category. Nevertheless, when only muscle foods (e.g., red meat, fish, poultry) are consumed, seafood-borne illness represents 56% of all outbreaks and 21% of all cases when incidents of unknown etiology are included.

Natural seafood toxins – mainly ciguatera and scombroid poisoning and, to a lesser extent, paralytic shellfish poisoning – were responsible for 62.5% of all seafood-borne outbreaks of illness, but constituted only 28% of all reported cases. Shellfish-related incidents, although responsible for only 31% of the seafood illness outbreaks, involved 66% of all seafood-borne cases. Most of these (55%) were registered as of unknown etiology but are believed to be due mainly to Norwalk, Norwalk-like, or human enteric virus infections, with a smaller proportion caused by *Vibrio* bacteria. Fish-borne incidents due to causes other than natural toxins were only 9% of all outbreaks and 8% of all cases. They resulted mainly from bacteria, including common food-borne disease organisms, and from unknown etiology, suspected to be primarily enteric virus or recontaminant vibrios. Botulism is a specialized but significant component of fish-borne disease. Disease due to parasites was minimal (0.4% of outbreaks and 0.6% of cases reported from seafoods).

Shellfish-borne disease occurs mostly from molluscs consumed raw or lightly heated, which constitutes the largest consumer risk. Ciguatera is a highly regionalized and intense risk for inhabitants and visitors consuming certain reef-associated fish in Caribbean and tropical Pacific islands and in adjacent mainland areas. Scombroid poisoning is widely distributed geographically but is specifically associated with consumption of certain fish species, particularly tuna, mackerel, mahimahi (dolphin), and bluefish. Botulism is a hazard for native American groups in Alaska that eat traditional fermented seafoods. Other risks are typical of food-borne disease in general and result from errors in handling, storage, or processing procedures. These are no greater than for other foods of animal origin.

Intolerance to eating certain types of seafood is rare and more typically associated with certain individuals in risk categories predisposed by other health complications. Seafood allergies, distinguished as immunological reactions rather than the inability to digest, appear to be more prevalent, but they are difficult to diagnose and document. Specific allergens in seafood have thus far been only grossly characterized in few studies. Seafood intolerances and allergies can be due to food additives (e.g., sulfites) that cause symptoms and confuse diagnoses. Additional investigation of the biochemical and immunological characteristics of seafood allergies and their significance seems warranted. In light of this level of information on the cause and occurrence of this somewhat limited form of seafood-borne illness, regulatory response must depend on proper labeling to distinguish (1) species or seafood type, (2) ingredients in formulated and fabricated seafoods [e.g., fish base surimi (a washed mince of the separated muscle tissue from fish to which cryoprotectants are added) formed to resemble crab], and (3) ingredients used in preservation and processing (e.g., sulfites to retard shrimp melanosis).

MICROORGANISMS AND PARASITES

Extent of Risk

Seafoods, like any food item, have the potential to cause disease from viral, bacterial, and parasitic microorganisms under certain circumstances. These agents are acquired from three sources: (1) mainly fecal pollution of the aquatic environment, (2) the natural aquatic environment, and (3) industry, retail, restaurant, or home processing and preparation. With the exception of foods consumed raw, however, the reported incidences of seafood-related microbial diseases are low.

Available data from CDC and from the Northeast Technical Support Unit (NETSU) of the Food and Drug Administration (FDA) for 1978-1987, as well as literature reports, suggest that the greatest numbers of seafood-associated illnesses are from raw molluscan shellfish harvested in waters contaminated with raw or poorly treated human sewage. The majority of these illnesses have unknown etiologies clinically suggestive of Norwalk and Norwalk-like agents that cause human viral gastroenteritis. Although these are the most common seafood-associated illnesses, they tend to be relatively mild with no associated mortality.

Except for Guam, naturally occurring marine *Vibrio* species are responsible for fewer reported cases of infections involving the consumption of raw molluscan shellfish, but certain species such as *V. vulnificus* can be associated with high mortality (>50%) in persons who are immunocompromised or who have underlying liver disease.

The microbiological risk associated with seafood other than raw molluscan shellfish is much lower and appears to result from recontamination or cross-contamination of cooked with raw products, or to contamination during preparation followed by time/temperature abuse (e.g., holding at warm temperature long enough for microbial growth or toxin production to occur). This occurs mainly at the food service (postprocessing) level, which is common to all foods and not specific for seafood products.

Seafood-related parasitic infections are even less common than bacterial and viral infections, with anisakids and cestodes having the greatest public health significance in the United States. In general, parasitic infections are concentrated in certain ethnic groups that favor the consumption of raw or partially cooked seafoods.

Thorough cooking of seafood products would virtually eliminate all microbial and parasitic pathogens. Individuals who choose to eat raw seafood should be educated about the potential risks involved and how to avoid or mitigate them. In particular, immunocompromised individuals and those with defective liver function should be warned never to eat raw shellfish.

The greatest risks from the consumption of raw molluscan shellfish could be minimized by research to develop valid human enteric virus indicators for the proper classification of shellfish growing waters; by implementing and maintaining proper treatment and disposal of sewage to avoid human enteric pathogen contamination of harvest areas; by efforts to identify and limit the number of pathogenic *Vibrio* species in shellfish; by developing new diagnostic methods and improved processing technologies; and by applying risk-based regulatory and control measures for potential microbial pathogens in raw molluscan shellfish.

Other seafood-associated risks can be reduced by proper application of a Hazard Analysis Critical Control Point (HACCP) system. This cannot be achieved by the visual or organoleptic inspection currently used for meat and poultry. Seafood inspection should be directed toward identification of microbiological risks to consumers and the effectiveness of methods to reduce or eliminate such risks. Additional studies are necessary to determine levels of particular microorganisms that constitute a risk and that can be used as a basis for microbial guidelines. This requires appropriate epidemiological research. Inspection requirements should apply to imported as well as domestic products.

Principal Conclusions

Most seafood-associated illness is reported from consumers of raw bivalve molluscs, and is due to unknown etiologies but is clinically suggestive of Norwalk-like viral gastroenteritis. The majority of incidents are due to consumption of shellfish from fecally polluted water. The disease is usually mild and self-resolving.

Naturally occurring marine *Vibrio* species are responsible for fewer reported cases of infection from the consumption of raw molluscan shellfish, but species such as *V. vulnificus* can be associated with high mortality in persons who are immunocompromised or who have underlying liver disease.

A lesser risk of microbial disease associated with other seafoods – resulting from recontamination or cross-contamination of cooked by raw product, or to contamination from other sources – is usually associated with time/temperature abuse. The etiologic agents most commonly involved, in order of reported frequency, are *V. parahaemolyticus* , hepatitis A, *Salmonella, Shigella, Clostridium perfringens*, and *C. botulinum* (mostly limited to Alaskan natives).

Seafood-related parasitic infections are even less common than bacterial and viral infections, with anisakids and cestodes having the greatest public health significance in the United States. In general, parasitic infections have resulted from consumption of raw or partially cooked fresh- and salt-water fish of particular species (e.g., whitefish, salmon).

Principal Recommendations

Consumers should be informed of the risks of eating raw seafoods, particularly molluscan shellfish. Individuals belonging to high-risk groups, such as cirrhotics, people with hemochromatosis, or immunosuppressed individuals, must not eat raw shellfish; this requires that health professionals be educated concerning hazards to high-risk individuals.

Adequate and proper treatment and disposal of sewage must be implemented to avoid contamination of harvest areas by human enteric pathogens.

Valid indicators for human pathogen contamination of growing waters must be developed. Seafood-borne infections by human enteric viruses in raw and improperly cooked molluscan shellfish could be decreased significantly by the development of valid growing water indicator(s) and of direct detection methodologies for enteric viruses.

Effective enforcement to eliminate recreational commercial and illegal harvesting and sale of molluscan shellfish from contaminated growing areas should be developed and adequately funded.

Means must be investigated and implemented to eliminate, or at least reduce, levels of potentially pathogenic *Vibrio* species in raw shellfish. This may necessitate restriction of harvest when water temperatures are high, rapid cool-down and continued chilling of products, and possibly irradiation of live shellstock and shucked products.

Consideration should be given to monitoring *Vibrio* counts in molluscan shellfish during warm months.

Because of the high risks associated with raw molluscan shellfish, the importation of live shellfish for raw consumption should not be permitted unless standards for the microbial quality of harvest waters and of harvest processing in the exporting country are fully equivalent to those in the United

States. Consumers should be advised to cook seafood sufficiently to destroy parasites and bacterial contaminants before consumption.

Control systems for microbiological hazards must include inspection techniques, preferably HACCP based, that specifically test for the hazard itself or for some condition that enhances or reduces hazard. Valid microbiological guidelines, established with an appropriate epidemiologic data base, are needed for seafood products.

Special attention should be addressed to ensure the safety of seafoods processed by newer techniques, such as *sous vide* and the use of controlled atmosphere packaging, that are potentially hazardous.

New or improved methodologies [e.g., enzyme-linked immunoabsorbent assay (ELISA), gene probe, polymerase chain reaction] should be developed that provide for the rapid identification and quantification of indicators, seafood-associated pathogens, and microbial toxins in seafoods and harvest waters.

NATURAL TOXINS

Extent of Risk

Incidents of illness due to naturally occurring seafood toxins reported to CDC in 1978-1987 were limited to ciguatera, scombroid fish poisoning, paralytic shellfish poisoning (PSP), and neurotoxic (brevetoxic) shellfish poisoning (NSP). Other intoxications, including puffer fish poisoning (PFP), were reported earlier; diarrhetic shellfish poisoning (DSP) and amnesic shellfish poisoning (ASP) are possible risks that should be anticipated.

Naturally, toxic fish and shellfish are not distinguishable from nontoxic animals by sensory inspection, and the toxins are not destroyed by normal cooking or processing. Except for scombroid fish poisoning, natural intoxications are both highly regional and species associated, and toxins are present in the fish or shellfish at the time of capture. Scombroid poisoning is due to histamine produced by bacteria multiplying on fish that are mishandled after capture; illnesses are widely reported from different states.

Ciguatera is a sometimes severe disease caused by consuming certain species of fish from tropical waters usually associated with islands or reefs. The disease is most common (endemic) in the Caribbean and Pacific islands, with some outbreaks in southern Florida and sporadic cases in other states caused by imported fish or tourist travel to endemic areas. Ciguatera was responsible for about half of all reported outbreaks of seafood intoxication in 1978-1987. The treatment is largely supportive, but mortality is low. At present, no effective control systems are in place for the prevention of ciguatera because a test for toxic fish is not generally available. Warnings and advisories concerning the hazards of ciguatera and the risks of consuming particular species of fish from ciguatera areas are issued by various states. Active control is proposed based on regulation of fishing for dangerous species, supported by testing suspect fish at dockside or on board the fishing vessel to detect and reject ciguatoxic fish. Increased education of the consuming public, sports fishers, and health professionals on the hazards and symptoms of ciguatera is also recommended.

During the same period, scombroid poisoning reportedly caused a similar number of outbreaks as ciguatera but was much more widespread in occurrence. Tuna, mahimahi, and bluefish were implicated as the major cause of scombroid poisoning in the United States. The disease is generally mild and self-resolving, and symptoms can be ameliorated by antihistamine drugs. Because the histamine that causes scombroid poisoning is produced after the fish have been caught, as a consequence of improper temperature control, the disease can be prevented by rapidly cooling fish after capture to 10°C or lower (e.g., 0°C if kept for an extended period) and holding them at or below this temperature at all times before cooking and eating. A HACCP-based system would control this poisoning for commercially handled fish, but the education of subsistence and recreational fishers is also necessary.

Paralytic shellfish poisoning was a minor cause of seafood-borne illness in 1978-1987, with only two deaths reported. This is a remarkable record in view of the annual occurrence of toxic situations among shellfish on both the East and the West coasts of the United States, which indicates that current control measures applied by coastal states are highly effective. However, the increasing occurrence of toxic dinoflagellate blooms and changing eating practices (e.g., eating whole scallops) among some sectors of the consuming public require increased surveillance and the development of more rapid and simple tests for toxic shellfish.

Although none of the other natural seafood intoxications, except for a single outbreak of NSP, have been reported recently by U.S. consumers, the potential for their occurrence either from domestically produced seafoods or from imports is real. Increased vigilance concerning imported products based on a requirement for certified nontoxicity is recommended. Moreover, both state and federal laboratories should be prepared to test for these "other" toxins, and procedures to deal with outbreaks should be in place.

Principal Conclusions

Natural toxin risks are highly regional or species associated.

Natural toxins are present in the environment and are not affected by procedures during or after harvest. The one exception is scombroid shellfish poisoning, which is due to postharvest mishandling.

Reliable, rapid tests for the natural toxins are either unavailable or not fully developed.

Although PSP is well controlled by state inspection systems and industry controls are in place for scombroid poisoning, there are no regulations for the control of ciguatera.

Recreational and subsistence fish eaters are at particular risk from natural toxins, and there is a lack of understanding by consumers of this risk.

Principal Recommendations

Control for natural toxins in the food chain should be at, or prior to, harvest, either by closures or by testing at the point of harvest.

Scombroid poisoning should be controlled primarily by rapid chilling on the vessel and by maintenance of refrigeration temperatures throughout distribution. Major emphasis should be placed on the development of rapid assays for each of the other natural toxins; without this, control is very difficult.

Primary regulatory authority should be at the state level, with funding, quality control, and specialized assistance from a federal seafood safety program.

Imported seafoods must be certified to be free of natural toxins through equivalency arrangements or more effective memoranda of understanding (MOUs) with exporters. An MOU refers to a formal agreement between a U.S. government agency (e.g., FDA) and another government agency (federal, state, local), or an informal agreement with a foreign government or other foreign institution.

Educational programs on the dangers of natural seafood toxins must be developed for recreational and subsistence fishers, and health providers must be given information to improve the identification and treatment of illness due to seafood toxins.

CHEMICAL RESIDUES

Extent of Risk

Fish and shellfish accumulate chemicals from the environment in which they live, but the extent of accumulation depends on such factors as geographic location, species of fish, feeding patterns, solubility and lipophilicity of the chemicals, and their persistence in the environment. Moreover, whereas land animals used for human consumption are fed mostly food of plant origin, aquatic animals that contribute to the human diet are generally predators of other animals and, in some cases, predators of predators. Because of this, chemicals have an opportunity to become more concentrated through bioaccumulation.

The most difficult area for risk evaluation is the problem of chemical residues because the health effects suspected do not take the form of obvious, distinctive, and acute illnesses. The potential risks of concern (e.g., modest changes in the overall risk of cancer; subtle impairments of neurological development in fetuses and children) are generally quite difficult to measure directly in people exposed at levels that are common for U.S. consumers. Immunoincompetence may increase cancer risk. Inferences about the potential magnitude of these problems must be based on the levels of specific chemicals present, on observations of human populations and experimental animals exposed at relatively high doses, and on reasonable theories about the likely mechanisms of action of specific toxicants and the population distributions of sensitivity and human exposure. In nearly all cases the current state of knowledge on these subjects must be regarded as quite tentative. Additionally, the number and variety of chemical residues are substantial, although a small minority constitute the bulk of the risk that can be assessed quantitatively at this time.

Overall, several chemical contaminants in some species of aquatic organisms in particular locations have the potential to pose hazards to public health that are great enough to warrant additional efforts at control. Available information suggests that these risks, in the aggregate, are not generally of a magnitude comparable to the highest environmental health hazards characterized to date; nevertheless, their control would significantly improve public health. Some examples of risks that may be significant include reproductive effects from polychlorinated biphenyls (PCBs) and methylmercury; carcinogenesis from selected congeners of PCBs, dioxins, and dibenzofurans (all of which appear to act primarily by binding to a single type of receptor); and, possibly, parkinsonism in the elderly from long-term mercury exposure. Several other metallic and pesticide residues also warrant attention.

Principal Conclusions

A small proportion of seafood is contaminated with appreciable concentrations of potentially hazardous organic and inorganic chemicals from both natural and human sources. Some examples of the risks that may be significant include reproductive effects from PCBs and methylmercury, and carcinogenesis from selected PCB congeners, dioxins, and chlorinated hydrocarbon pesticides.

Consumption of some types of contaminated seafood poses enough risk that efforts toward evaluation, education, and control of that risk must be improved. Present quantitative risk assessment procedures used by government agencies should be improved and extended to noncancer effects.

Current contaminant monitoring and surveillance programs provide an inadequate representation of the presence of contaminants in edible portions of domestic and imported seafood, resulting in serious difficulties in assessing both risks and specific opportunities for control.

Due to the unevenness of contamination among species and geographic sources, it is feasible to narrowly target control efforts and still achieve meaningful reductions in exposure.

The data base for evaluating the safety of certain chemicals that find their way into seafood via aquaculture and processing is too weak to support a conclusion that these products are being effectively controlled.

Principal Recommendations

Existing regulations to minimize chemical and biological contamination of the aquatic environment should be strengthened and enforced.

Existing FDA and state regulations should be strengthened and enforced to reduce the human consumption of aquatic organisms with relatively high contaminant levels (e.g., certain species from the Great Lakes with high PCB levels, swordfish and other species with high methylmercury levels).

Federal agencies should actively support research to determine actual risks from the consumption of contaminants associated with seafood and to develop specific approaches for decreasing these risks. Increased environmental monitoring should be initiated at the state level as part of an overall federal exposure management system.

States should continue to be responsible for site closures, and for issuing health and contamination advisories tailored to the specific consumption habits, reproductive or other special risks, and information sources of specific groups of consumers.

Public education on specific chemical contaminant hazards should be expanded by government agencies and the health professions. For specific contaminants in particular species from high-risk domestic or foreign geographic areas, government agencies should consider the option of mandatory labeling.

Additional study of potential chemical contamination risks associated with both domestic and imported aquaculture products is required. Because of different standards for drug or agricultural chemical use and water quality prevailing in other countries, imported aquaculture products should be effectively certified as meeting U.S. standards.

SCOPE AND ADEQUACY OF CURRENT SEAFOOD SAFETY PROGRAMS

Regulatory Guidelines, Monitoring, and Inspection

The current system of governance designed to protect the U.S. seafood consumer is composed of an intricate and complementary system of programs at the federal and state levels of government. Additional programs have been instituted in the private sector that offer a measure of industry self-regulation. At the federal level the principal responsibility for setting regulatory guidelines and for the surveillance and control of seafood safety is divided among the FDA, the Environmental Protection Agency (EPA), and the National Marine Fisheries Service (NMFS).

Within states, responsibility may lie with one or more of their health, environmental, fishery, or agricultural departments. States generally tend to adopt federal regulatory guidelines.

A primary role for the federal government is setting regulatory guidelines designed to promote inspection and enforcement activities both within and outside formal governmental programs. Existing regulatory guidelines can be divided into (1) those designed to reduce acute risk from microbial and natural toxin contaminants, and (2) those designed to reduce long-term or chronic risk due to chemical contamination. Guidelines for microorganisms and natural toxins are determined solely by the FDA and have been set primarily on an as-needed basis, that is, in response to a reported public health problem.

Properly collated and effectively presented guidelines could provide a strong basis for the production and supply of safe seafood. However, in several areas related to new processing techniques and other emerging problems, new guidelines seem both appropriate and necessary. Setting federal guidelines for residual chemical contaminants is a task shared by EPA and FDA. Their strategy has been to focus on a limited number of chemical contaminants and to set regulatory limits by means of "action levels." Results of various federal and state efforts to monitor contaminant loads in the nation's marine and freshwater environments suggest strongly that several chemicals require a more fundamental review and evaluation.

In terms of assessing and managing risks, the overall posture of relevant federal agencies, particularly FDA, appears to be almost totally reactive. In the committee's judgment, there has been less effort than would be desirable to discover and quantify hazards that are not yet on the public agenda, to evaluate options for reducing risks, and to implement policies that protect both the health of consumers and the stability of commercial markets.

One of the more important activities at both the federal and the state levels is environmental monitoring. Because the majority of seafood is from wild stocks, the quality of harvesting waters is of fundamental concern. The EPA and certain state governments [primarily by way of their involvement in the National Shellfish Sanitation Program (NSSP)] have instituted programs to establish the level of contaminants in seafood harvesting waters.

These efforts have led to important insights into general water quality but, for the most part, do not supply sufficient information on the question of seafood safety. Among other things, they lack (1) sufficient geographic scope, (2) a common methodological approach, and (3) sufficient focus on the edible portion of seafood in order to determine public health, as opposed to environmental health, impacts. This last point is an important one. Except for the monitoring of harvesting waters carried out as part of the NSSP, data evaluating contaminant levels in fish and shellfish do not consistently focus on the analysis of edible tissue. More often the focus is on whole fish or on liver and gallbladder analysis. These analyses, by their design, offer insufficient insight into contaminant levels in the edible portion of seafood products.

Inspection efforts by FDA and various state and local public health agencies are designed to ensure safety, but are insufficient to ensure in all cases that the regulatory guidelines defined by FDA and EPA are not being exceeded. The sampling strategies employed by these various agencies are designed to focus inspection and enforcement activities on areas in which the probability of a problem appears highest. Ongoing governmental efforts to develop new inspection programs, with a focus on the public health aspects of the raw product and the environment from which these products are derived, along with continued control of seafood production and processing, could provide measurable additional benefits in seafood safety.

Given many of the intrinsic attributes of seafood already discussed, it is clear that an approach recognizing the advantages of regional/local control and surveillance is essential. The question of seafood safety should continue to be one in which federal and state roles are viewed as a cooperative partnership. It is also apparent that seafood commerce is taking place within an increasingly interdependent international economy. Many of the major trading partners of the United States are developing or further refining formal regulatory programs for seafood safety. These efforts should be taken into account in designing a domestic program.

Principal Conclusions

Federal (mostly FDA) guidelines for microbial and natural toxin contamination should be extended and updated. Those that exist have not been adequately conveyed to the fishing industry and to interested members of the public.

Federal guidelines on chemical contaminants in seafoods are limited in scope and, in some cases, questionable as to the levels set. There is an apparent lack of coordination in the development and use of data on chemicals in the aquatic environment among FDA, EPA, the National Oceanic and

Atmospheric Administration (NOAA), and the states. Better recognition is required of the importance of regional factors in the occurrence of toxic fish and shellfish and of the existence of high atrisk groups (e.g., pregnant women, children, recreational and subsistence fishers).

The present federal monitoring and inspection system is too limited in frequency and direction to ensure enhanced safety of seafoods. The monitoring process depends too much on evaluation of the product, rather than on safety of raw materials, with the single notable exception of the NSSP. However, even NSSP is not providing adequate protection because molluscan shellfish appear to cause most seafood-borne disease.

Recreational and subsistence fishing is largely ignored in health and safety monitoring at the federal level. Consumers of seafood from these sources can be at high risk from natural toxins and chemical pollutants in certain regions and in particular species of fish. The health risks include cancer and the subtle impairment of neurological development in fetuses and children.

The present system of data collection on seafood-borne illness by CDC does not provide an adequate picture of the extent and causes of such disease.

Seafood advisories warning of local or species-associated health risks are issued mostly by state authorities and vary greatly in both their content and their distribution. Nevertheless, these advisories serve a useful purpose.

Because of the regional nature of much of the domestic fisheries problems, states seem the logical level at which to tackle seafood control problems. However, help and guidance from the federal level are required.

State programs for monitoring, surveillance, and control of seafood safety are generally in place in coastal states that use federal guidelines and action levels where these are available. However, the quality and effectiveness of the programs vary greatly as a function of the financial and administrative support available to the responsible state units, and in accordance with the character of the resource. A greater emphasis should be placed on the development of formal arrangements with foreign producers to guarantee that imported seafood has been harvested and processed in noncontaminated environments.

Present training and education of industry and regulatory personnel are too limited both in scope and in number. Insufficient attention is given to the education of physicians and other health professionals on seafood safety and the characteristics of seafood-borne disease. This is also true of the consuming public.

The regulation of imported seafoods to ensure safety is largely based on end product inspection and testing, except where MOUs exist. This is ineffective because it involves a mainly reactive process.

The regulation of imported seafood products is carried out largely without regard to other national or international programs. There is tremendous variance in both regulatory limits for contaminants and inspection protocols in various countries, which leads to excessive and cumbersome inspection strategies for the importing state, and may also lead to a general reduction in the number of countries engaged in international seafood trade in the future.

Principal Recommendations

A more concise, comprehensive, and generally available single source for all FDA guidelines relating to seafood safety should be developed and updated on a regular basis. This information should be disseminated to industry and integrated into state regulatory processes through more routine and uniform training programs.

The development of an interagency structure with a single focus on seafood safety could contribute significantly toward increasing communication within the federal regulatory system, but the responsibility for primary control should be with the state.

Federal agencies should develop a set of monitoring and inspection practices focusing more strongly on environmental conditions and on contaminant levels in the edible portion of seafood at the point of capture. Strong consideration should be given to creating a marine recreational fishing license system that is linked to the distribution of information characterizing the level and scope of potential risk from eating recreationally caught fish. Strong consideration should also be given to the closure of recreational harvest areas deemed to pose a threat to human health.

The CDC should develop an active and aggressive program, founded on community-based health surveys, to better determine the level and source of seafood-borne illness in the U.S. population.

Consideration should be given to the development of agreements with foreign authorities and individual producers to ensure that imported products are treated in a manner consistent with and equivalent to domestic products.

A more pronounced and consistently defined federal role in the risk characterizations leading to seafood health advisories should be developed. A more consistent and focused effort in determining and communicating public health risks from contaminated seafood should also be developed.

As more countries require the equivalency of domestic and imported products, it is apparent that the time has come for the international community to begin a process that would minimize the differences existing among national regulatory guidelines and approaches.

OPTIONS FOR REDUCING PUBLIC HEALTH RISKS

Monitoring, Control, and Surveillance Measures

The current system involves (1) surveillance by federal and state agencies to identify seafood-borne disease (e.g., CDC and state health departments); (2) evaluation of risk and setting of guidelines and action levels mostly by federal agencies (e.g., EPA and FDA); (3) control of risk by inspection and testing of edible fish and shellfish (e.g., states, FDA, and NMFS); and (4) action to protect consumers by embargo, detention, seizure, or recall, and by issuance of warning advisories (e.g., states and FDA). This system needs revision and strengthening to develop a truly risk-based regulatory process.

The data base on which regulation depends is inadequate. The disease surveillance system of CDC suffers from inadequate resources and should be refocused to provide a more complete and balanced account of seafood-borne disease. More analytical data on contaminants are needed, which could be obtained by increasing FDA analyses and sponsoring broader integrated studies of marine and fresh waters by EPA and corresponding state agencies.

Inspection and testing should focus on actual problems (as in HACCP systems), and there should be increased efforts to develop rapid, reliable test methods for dangerous microorganisms, toxins, and contaminants. This will require a restructuring of inspectional systems to accommodate newer methodologies and to train personnel in their application. Emphasis on purely sensory evaluation should be decreased.

Problems of interagency jurisdiction, unclear regulations, or poor cooperation among state and federal agencies should be addressed and rectified. This will require added resources.

Characteristics of Control Requirements

Control measures should be applied initially at the earliest stage of seafood production by monitoring of water quality and condition. Such measures would apply to the molluscan shellfish problem and to most natural toxins and chemical contaminants, and would permit the exclusion of potentially dangerous fish or shellfish from markets by fishing closures and use of advisories. Rapid and simple tests should be developed and used to screen potentially hazardous fish or shellfish at the point of harvest to reduce costs to the fishermen and to protect the consumer from toxins and dangerous contaminants. Postharvest control seems likely to be most readily achieved through an HACCP-based system focusing on cross-contamination, temperature control, and the effectiveness of handling and processing methods designed to inhibit or destroy microorganisms. This system must be based on safety considerations, not solely on quality.

The extent of chemical contamination of seafood species is both species and region dependent. A few chemicals such as mercury have strong species associations (e.g., swordfish). The concentrations of most organic chemicals tend to be less species associated and more dependent on geographic region. Within aquatic organisms, bioaccumulation may be organ specific or related to fat concentrations (e.g., methylmercury in muscle tissues and PCBs in fatty tissues).

Improvement of the total data base on chemical contamination of fish could enable regulatory agencies to target their efforts on particular species of fish in specifically defined areas and, thus, lead to considerable mitigation of individual and societal health risks with minimal economic effects. Improvement of the data base could also enable consumers, especially subsistence and sport fishers, to select the least toxic fish in their waters for consumption. Clearly, however, chemical contamination is ultimately a problem of environmental degradation due to waste dumping that can be solved only by the development of systems to reduce chemical disposal in fresh and marine waters and in the atmosphere. The improvement of environmental quality will mean safer fish.

The effectiveness of current fishing controls and consumer/fisher information programs in geographic areas with greater-than-average contamination problems is uncertain. Unfortunately, contaminated areas may be pocketed within broader fishing grounds, and the precise distribution of relatively high residue levels may be difficult to determine. In areas such as the Great Lakes, steps have been taken to prevent the commercial distribution of fish that have contaminants exceeding established tolerance or action levels. However, the adequacy of some regulatory levels is open to question in light of newly available information, and the degree of protection afforded the substantial population of consumers of sport caught fish by advisories based on those tolerance/action levels is even more doubtful.

Better control is needed of imported fish products, which represent over half of the fish and shellfish consumed in the United States. Seafood imports are coming from an increasing number of countries, some of which have poor internal control systems. A significant number of supplier countries are in tropical areas where some bacteria and toxin hazards are intrinsically high. Additionally, the United States is receiving increasing numbers of fish and crustaceans from foreign aquaculture operations. In view of the often regional and species-associated nature of seafood hazards, it would be appropriate to classify suppliers into risk categories for particular species or processed seafoods.

Consideration should be given to extending the scope of MOU arrangements to cover all seafoods, and unfettered import of seafood products should be permitted only from countries with whom the United States has MOUs. Testing of imports for chemical residues should be carried out systematically according to a planned program designed to provide long-term estimates of the level of contamination in particular species or in the products of different supplier countries. In view of the very complex structure of international trade in fishery products, it is desirable that a better system be established to identify the country of origin of imported seafoods. This may require international agreements.

LEGISLATIVE CONSIDERATIONS

Education and Information Measures

Programs should be established for training regulators and seafood industry personnel to be proficient in the regulatory programs under consideration. These programs should be well-coordinated across states, with more national guidance and increased consideration of the unique attributes of various geographic regions.

States should be required to produce advisories that can be used by both commercial and recreational personnel to learn about local public health risks and protective measures. In the development of advisories for reproductive effects, due weight must be given to the persistence of different toxicants in people. A useful federal function, besides producing advisories to meet national problems, would be development–with the states–of a standardized format for written and broadcast advisories so that there will be minimum confusion due to state and local differences.

Educational programs for safe preparation and service of seafoods in commercial and home settings must also be developed and delivered as a part of an integrated seafood safety program.

Recommendations for Improved Inspection Strategies

Inspection should continue to be based on shared responsibility between state and federal agencies.

The general philosophy presented here involves the concept of a federal agency (or agencies) having responsibility for identifying and characterizing risks, establishing methodologies and acceptable or actionable levels of undesirable agents, and monitoring state inspection programs.

In addition, the federal agency would continue to have primary responsibility for imported products and products in interstate shipment. The agency would establish well-equipped regional laboratories to conduct tests for the federal program and–where appropriate–for state agency programs.

States, with the financial support of federal sources, would carry out inspections and apply police powers to in-state fishing industry operations, by using methods and procedures that meet or exceed federal standards.

Monitoring and certifying state programs to determine their eligibility for federal funding of such operations would be the responsibility of the federal agency. Results would include better use of state agencies and strengthening their role in inspection, rather than depending solely on a federal agency to perform nationwide inspection.

The federal agency should have overall responsibility for coordinating the national program and carrying out those functions that states cannot or will not undertake, as well as ensuring the training of state personnel. Organoleptic inspection must be recognized as inadequate and of little value for seafood safety because it is unable to identify risks to humans.

Where new legislation is being considered in relation to the problems of seafood safety, the following important points should be considered:

(1) the need to facilitate closure of harvesting areas on the basis of human health hazards,

(2) the need for a strong state role in inspection that will require federal support,

(3) the desirability of regulating vessels and dock facilities in relation to human health,

(4) the collation of current regulations in easily available form, and

(5) the need to train state and federal regulatory personnel.

POTENTIAL IMPACT OF PROPOSED OPTIONS

The proposed options outlined above will have the following impact on seafood and the consumer:

(1) they will improve the general health of the public by focusing on the cause of disease, thus reducing the cases of seafood-borne diseases;

(2) they will produce a quicker, more effective response when the public is subjected to unacceptable risks;

(3) they will promote compliance through increased and improved communication among the involved agencies and industry, and through increased public knowledge; and

(4) they will require the appropriation of funds to develop a comprehensive system incorporating the above recommendations.

DIRECTION FOR DATA COLLECTION AND FUTURE RESEARCH

Inasmuch as accurate risk identification is the first step in risk-based control programs, stronger epidemiologic data are needed to assess the extent of public health risk in terms of incidence, severity, vehicle, and setting.

The two major viable data bases for seafood-borne illnesses from CDC and NETSU are too limited in scope and have discrepancies related to methods of surveillance and reporting that prevent consistent correlation of the outbreaks of some pathogens. In addition, more basic research is necessary to understand why and how certain pathogens or toxins cause illness. For example, there are bacterial pathogens, such as *Vibrio vulnificus*, or non-O1 *V. cholerae* that are commonly isolated from shellfish, that cause only a small number of clinical cases; we need to understand why only a minority of persons become ill after exposure to these organisms. Similarly, a better understanding is required of how natural toxins and chemicals are processed by fish, so that we can better predict when and where human illness will occur.

Rapid, nondestructive, and easy-to-perform tests for toxins, microorganisms, and chemicals [e.g., stick test for ciguatera, deoxyribonucleic acid (DNA) probes for specific viruses and bacteria, and instrumental chemical analysis] must be developed. The current programs for testing water and seafoods for potentially dangerous chemicals should be broadened to provide a satisfactory data base for regulation and control.

4

Aquatic Genetic Resources

Unlike agricultural crops, aquatic life largely inhabits publicly owned territory. Authority to manage fish genetic resources (which are still almost universally thought of as 'fisheries') may be divided between different levels of government, as in Canada, where the federal government manages ocean resources as well as fish that migrate inland to spawn, while provincial governments are in charge of inland fish. In some cases, management authority may also be devolved to indigenous groups such as the Nunavut people of northern Canada, whose land agreement with the federal government gives them the right to control access to biological and genetic resources.

Depending on national constitutions, laws and administrative practices, there may be a chain of management authority that includes national governments, state/provincial governments, local governments, unorganized indigenous and local communities, and local cooperatives such as those that hold exclusive fishing rights in the Philippines. The concept of fish being caught for food is straightforward enough, but when fish take on the grander identity of 'aquatic genetic resources', everything changes.

Under that guise, they're being used as broodstock for aquaculture operations, or perhaps their DNA is being transplanted into a different species. In those circumstances, it might be said, a fish is no longer a fish but a collection of 'genetic material' – or, as the CBD puts it, 'material . . . containing functional units of heredity' (eg sperm, eggs or DNA). When fish are used as aquatic genetic resources, ownership takes on a whole different level of importance – both because that fish may be worth much more as a genetic resource than as a barbecued fillet and because the knowledge of how to use the fish as a genetic resource has its own separate worth– more on this later in the chapter. The bottom line is that the multiplicity of fisheries management regimes in different countries can't do the job of determining who has the right to collect – or permit the collection of – an aquatic genetic resource. It doesn't help matters that it may not always be easy to tell whether a fish is being used simply as a fish (eg a cardinal tetra for display in an aquarium) or as a genetic resource (a tetra for breeding), as our earlier discussion of the ornamental fish trade in the Rio Negro (Case Study 1) made clear.

For most of the 20th century, aquatic genetic resources were considered common property, available for everyone's use regardless of the location of the resource or the nationality of the user. Two international agreements in the early 1990s radically changed the rules for ownership and control of aquatic life. These were the UNCLOS and the CBD.

OCEAN RESOURCES: THE LAW OF THE SEA

The Law of the Sea came into force in 1994 and has been signed by more than 130 countries. An international agreement that sets conditions and limits on the use the oceans, including the seabed, UNCLOS establishes the rights and obligations of coastal states in contiguous waters, sets conditions for the conduct of marine research, and obliges countries to provide access to surplus catch.

Sovereignty

UNCLOS both defines and limits coastal countries' sovereignty over the seas and their biodiversity. Coastal states can claim full sovereignty over their territorial sea to a maximum of 12 nautical miles (22.2 km) (Article 3). Beyond the territorial sea, UNCLOS creates a 200 mile Exclusive Economic Zone (EEZ) in which coastal countries can claim exclusive rights for exploration, exploitation, conservation, and management of all natural resources (living or non-living) on the seabed and in its subsoil, and overlying waters (Article 56). Some countries have not yet ratified UNCLOS, and many have not yet claimed their EEZ. Beyond the EEZs are the high seas, to which all countries have open access rights. The CBD, in Article 22, confirms the rights and obligations established under UNCLOS and requires all parties to implement the convention consistently with those rights and obligations.

Research

UNCLOS grants each coastal state the exclusive right to regulate, authorize, and conduct scientific research in its territorial sea (Article 245). However, each state is required to consent, in normal circumstances, to scientific research projects in EEZs or on the continental shelf (which includes the territorial sea) by other states or international organizations 'to increase scientific knowledge of the marine environment for the benefit of all humankind' (Article 246). In turn, foreign researchers must provide information on their research activities, accept the participation of the coastal state research programmes, and allow the coastal state access to all data and samples derived from the marine scientific research project (Article 249). These provisions, designed to facilitate access and benefit sharing at the national level, apply to waters adjacent to coastal communities, as the territorial sea begins at the low water line.

Surplus Catch

Each state must determine its capacity to harvest living resources within its EEZ. If it is unable to meet the allowable catch it has established, it must provide other states access to the surplus (Article 62). While this provision is clearly intended to apply to food fisheries, it could theoretically apply as well to other life such as seabed invertebrates.

INLAND WATERS: NATIONAL SOVEREIGNTY

Biological diversity flourishes most richly in tropical regions while the countries with the technology to make the most advanced use of genetic resources are generally the northern, most developed countries. Consequently, the benefits from the commercial use of genetic resources have largely been enjoyed by companies in the North. Yet, following the new found global concern about disappearing biodiversity, the responsibility for conservation inevitably falls on the poorer southern countries. The rush for genetic resources by transnational corporations has been unfavorably compared to Europe's colonization of southern countries, two centuries earlier, to gain access to the raw materials needed for the Industrial Revolution (eg Shiva, 1997).

During the negotiations leading to the CBD, some southern countries insisted on recognition of national sovereignty over genetic resources as a condition of agreeing to the CBD. In theory, recognition of national sovereignty meant the ability to control access and negotiate a fair share of the benefits arising from the use of genetic resources.

The CBD is the most comprehensive international instrument for the management of biodiversity. Its three objectives are the conservation of biological diversity, the sustainable use of its components, and the fair and equitable sharing of the benefits arising out of the utilization of genetic resources. The extent of global commitment to the CBD is reflected by the fact that, by June 2004, it had been ratified by 188 countries – almost all of the nation states in the world (with the notable exception of the US). While the Convention doesn't have the force of law, ratification signifies a commitment to enact national laws to bring its provisions into effect.

In addition to recognizing national sovereignty, the CBD creates rights and responsibilities for member countries regarding access to genetic resources. It calls on all Parties to 'create conditions to facilitate access to genetic resources for environmentally sound uses' by other Parties. It requires provider countries' informed consent to access to genetic resources and calls on collecting parties to share with provider countries 'in a fair and equitable way' the results of research and development and the benefits arising from commercial and other uses.

Recognition of national sovereignty under the CBD means that aquatic genetic resources in the wild are owned by the country in which they are located. National sovereignty over public lands extends as well to inland waters (with rare exceptions, such as in some Scandinavian countries) and to the aquatic life they contain. Although the CBD recognizes sovereign rights over genetic resources, it must be borne in mind that tenure and ownership systems are neither uniform across all countries nor clearly defined in any given country. Ownership rights may range from traditional common tenure to state enforced private rights to land and natural resources, including biodiversity (Columbia University, 1999). The fact that aquatic genetic resources are more likely than plant genetic resources to be publicly owned has significant implications for the management of access. One of these is that the resources commonly exist within the territories of traditional communities that may have rights to control their use.

The Implications of National Sovereignty for Conservation and Aquaculture

National sovereignty over genetic resources is clearly a done deal, but whether it will advance the cause of conservation is debatable. Certainly that wasn't the motivation of the developing countries that insisted on recognition of national sovereignty in return for signing the CBD. They were more concerned with having a substantial say in the burgeoning international trade in genetic resources – not that they could be faulted for resentment over past exploitation by northern countries and corporations. Nevertheless, for the CBD's conservation objective to succeed, both developing and developed countries will need to come to the realization that it's in their own best interests to prevent the further erosion of genetic diversity.

It was not without good reason that the FAO International Undertaking formally adopted the concept of genetic resources as the common heritage of humankind. The reasoning of that agreement was that the principle of free exchange was essential for the most beneficial exploration and use of genetic diversity for plant breeding and other scientific purposes. As Jain (1994) points out, as the world's population continues to grow rapidly, it will become increasingly important to identify new genetic variability for higher crop yields. This will require a great deal of international cooperation in research, including the ready exchange of plant genetic resources and new technologies.

If that's true for plant genetic resources, it will also be so for aquatic genetic resources during the next two decades. Developing countries went from consuming 45 per cent of the world's fish to 70 per cent today. It is expected that by 2020 that figure will increase to close to 80 per cent (IFPRI, 2003). Given the declines in wild fish populations, it appears likely that capture fisheries will be unable to meet the demand and that the gap will need to be filled by a growth in farmed production. Existing aquaculture operations have barely tapped the possibilities offered by genetic diversity among food fish species.

As discussed, the efficiency with which the aquaculture industry is able to meet global food needs will depend to a large degree on effective international cooperation in sharing genetic material and knowledge. As national access laws are still in the formative stages and are being written with an almost exclusive focus on plants, it's too early to say how big an impact national sovereignty (and, for that matter, emerging community rights of consent) will have on international sharing of fish genetic resources. It would be both ironic and unfortunate if the door to best quality fish genetic resources was to be shut at a time when aquaculture is poised to overcome its environmental shortcomings and begin to make a major contribution to global food security.

OWNERSHIP IN TRADITIONAL COMMUNITIES

National systems for ownership of genetic resources are often very much at odds with traditional community practices. Most traditional communities in developing countries have continued to apply their own tenure system for biological resources, while the state enforces private and public property rights on goods and resources, and intellectual property laws on industry and commerce. Some traditional tenure systems regarding genetic resources are grounded on collective ownership or heritage, and sometimes, particularly in the case of medicinal plants, on religious and mystical considerations (Khalil, 1995).

In many cases where governments have asserted sovereign rights over genetic resources, the land remains under traditional community tenure.

The end result is that the state reaps the benefits from agreements for the use of resources, while the burden of conservation might be said to rest with the communities. The danger is that, once communities have lost control over local resources as well as the ability to practise traditional livelihoods, they may lose as well both the knowledge and the motivation to take care of local ecosystems. The choice between doing whatever it takes to feed one's children and conserving an endangered fish population is no choice at all.

Although the CBD's provisions apply to agreements at the national level, they also take into account the role of traditional communities in conserving and using biodiversity.

Article 8(j) requires Parties to the CBD to 'respect, preserve and maintain knowledge, innovations and practices of indigenous and local communities embodying traditional lifestyles relevant to the conservation and sustainable use of biological diversity' and to encourage the equitable sharing of benefits arising from the use of such knowledge, innovations and practices. Article 8(j) confers no rights to communities but is based on the premise that sharing in benefits from the use of genetic resources will help contribute to their conservation and sustainable use.

How to implement Article 8(j) has been one of the greatest challenges for CBD Parties. Discussions at the Conferences of the Parties have generally assumed (a) that the use of genetic resources generally depends on traditional knowledge and (b) that those using genetic resources are usually based in countries other than the provider country.

As we'll explain in the following chapter, the first, critical point is far less true for aquatic than for plant genetic resources. Indigenous peoples, local communities and their advocates have high hopes that Article 8(j) will mean greater recognition of community rights over traditional resources, but this may not be the result for fishing communities if these rights are tied to an intimate knowledge of genetic resources.

RECOGNITION OF INDIGENOUS PEOPLES' RIGHTS

In the traditional cultures of indigenous peoples in widely separated parts of the world, the concept of 'ownership' has little meaning. First Nations participants at a workshop we hosted during the preparation of this book emphasized that they consider it inappropriate to refer to fish as 'resources' – genetic or otherwise. Indigenous peoples' attitudes about equality or 'oneness' of all living creatures, while they may be an essential part of spiritual beliefs, are likely to be rooted in a common sense recognition that surviving in a natural ecosystem requires maintaining the cycles of species, and that ownership is superfluous where there is abundance to be shared.

But times change. Today, claims of ownership make a frequent appearance in the rhetoric of the indigenous rights movement – but mainly as matter of political necessity in campaigns to win legal recognition of the right to use lands and biodiversity in traditional areas of occupation. 'Ownership' and 'title' are words well understood by courts that may have more difficulty relating spiritual beliefs to Western property law.

Ownership of land and control over aquatic life has become a contentious issue in coastal and riparian indigenous communities because of the broad extent of and fragility of fish habitat, the traditional importance of fish for sustenance, and frequent uncertainty about rights of access to fisheries. Recent attention to fish as genetic resources has added a new twist that isn't an issue for plant genetic resources: in the case of migratory fish, several indigenous communities may claim traditional ownership and the right to be consulted by a collector regardless of where broodstock is collected if a fish passes through their territory at some stage in its life cycle.

Most countries have made little progress towards resolving indigenous rights issues. This will undoubtedly impede efforts to facilitate access to genetic resources in indigenous communities. One notable exception is the Philippines, which under the 1997 Indigenous Peoples Rights Act extends a prior informed consent requirement to virtually all potential uses of natural resources in recognized indigenous territories. Another is Canada, where court decisions recognizing indigenous rights have led government negotiations of treaties with some indigenous peoples. Case Study 3 at the end of this chapter, although it describes an unsuccessful negotiation with indigenous communities for access to broodstock, illustrates one useful model for access negotiations with communities with co-management or ownership rights.

Although there is no unanimity among indigenous groups about the approach proposed by the CBD – some have turned their backs on the Convention, while others are actively working with national governments to develop workable access laws – there is naturally widespread scepticism about how Article 8(j) will be interpreted to ensure fair treatment and meaningful benefits for communities.

In 1993, the Draft United Nations Declaration on the Rights of Indigenous Peoples confirmed their right to 'own, develop, control and use the land and territories, including the total environment of the lands, air, waters, coastal seas, sea ice, flora and fauna and other resources which they have traditionally owned or

The CBD's recognition of national sovereignty over genetic resources represented an about-face from another international agreement more than a decade earlier. In 1981, member countries of the UN FAO adopted the International Undertaking on Plant Genetic Resources as an instrument to promote international harmony in matters regarding access to plant genetic resources for food and agriculture. The purpose of the Undertaking, to be monitored by the Commission on Genetic Resources for Food and Agriculture (CGFRA), was to 'ensure that plant genetic resources of economic and/or social interest, particularly for agriculture, will be explored, preserved, evaluated and made available for plant breeding and scientific purposes' (CGFRA, 2002).

The International Undertaking's recognition of genetic resources as the 'common heritage of mankind' was in large part a response to a demand from developing nations to keep plant genetic resources in the public domain and to put a brake on further privatization of agricultural genetic resources. Their concerns included gene banking in northern countries of seeds produced by southern farmers; the accumulation of seed companies under the control of transnational corporations; and the implementation of monopoly Plant Breeders' Rights (essentially a patent) over crop varieties, limiting farmers' access to breeders' lines and finished varieties.

Conflict over access to plant breeders' lines in the years following the approval of the International Undertaking led to a 1989 annex recognizing farmers' rights as a counterbalance to plant breeders' rights. Among other things, farmers' rights established the rights of farmers and their communities to participate fully in the benefits derived from plant genetic resources. By the early 1990s, as the privatization of agricultural research and patenting of plant genetic resources in industrial countries increased, developing countries were expressing disillusionment about the effectiveness of the International Undertaking and promises to implement farmers' rights. Consequently they proposed replacing the concepts of common heritage and open access with national sovereignty and benefit sharing to allow nation states to better control and benefit from their biological resources (GRAIN, 2000a). The enshrinement of these principles in the CBD necessitated a renegotiation of the International Undertaking to ensure harmony between the two agreements.

The result was the International Treaty on Plant Genetic Resources for Food and Agriculture, approved by FAO member countries in 2001 after seven years of difficult negotiation. The treaty is intended to meet the needs of both plant breeders and farmers, guarantee the future availability of plant genetic resources for food and agriculture, and ensure fair and equitable sharing of benefits. The agreement, which covers 64 food crops accounting for 85 per cent of global human nutrition, places some constraints on intellectual property over the seeds in the multilateral system and imposes obligations for benefit sharing when accessed seeds are commercialized. Civil society organizations such as Genetic Resources Action International (GRAIN) and the Erosion, Technology and Concentration Group (ETC), while acknowledging the treaty as a step forward, also criticize it for not guaranteeing farmers' rights and doing too little to ensure equity and benefit sharing. The history of the International Undertaking illustrates the increasing tension between developed and developing countries over the expansion of intellectual property rights and the demands of developing countries and farming communities for greater control over their genetic resources and the use of traditional knowledge developed by farmers. It also underscores the almost exclusive emphasis on plant genetic resources in international negotiations on access and benefit sharing. The International Undertaking and the treaty that succeeded it were made necessary by the interdependence among countries for access to crop germplasm needed to ensure global food security.

While there is currently no parallel interdependence in access to fish germplasm, the expanding importance of aquaculture in meeting global food needs may ultimately create the need to address similar issues to those that have dominated the negotiations towards the revised International Undertaking– most notably the compatibility between intellectual property rights protection and the protection of communities providing aquatic genetic resources for international aquaculture development.

otherwise occupied or used'. In 2001, the International Indigenous Forum on Biodiversity (IIFB, 2001) told the CBD Working Group on Access and Benefit Sharing that the 1993 UN Declaration represented the minimum acceptable standards and that the trust needed to meet the CBD objective of access, and that benefit sharing wouldn't come without recognition of the rights of indigenous peoples. How far to go in that direction remains a challenge for countries intent on facilitating access to genetic resources. Some indigenous groups see the CBD encouragement of benefit sharing with communities as a way to avoid tackling the hard issues that come with recognition of indigenous rights, especially if benefit sharing is limited only to those communities that can prove genetic resources users need their knowledge – and if benefits are solely tied to conservation objectives. The growing litany of 'biopiracy' complaints suggests that countries may find it increasingly hard to separate access issues from indigenous rights issues.

INTELLECTUAL PROPERTY LAW AND AQUATIC GENETIC RESOURCES

Patent applications have caused no end of controversy in the use of plant genetic resources. Naturally, a seed company that develops a genetically modified strain of millet resistant to a fungal infection, for example, considers it a top priority to protect its invention so that competitors don't copy it and flood the market with a cheaper version. But what if the company came up with the idea by modelling its invention on another strain developed by farmers in a traditional agricultural community? The last few years have seen a succession of international disputes – a notable example being the outcry that arose when a US company attempted to patent Basmati rice despite its long history in south Asia. What if a pharmaceutical company succeeds in developing and patenting a pain-killing drug after hiring a researcher to visit shamans in the Amazon, learn about their healing practices, and bring back samples of their medicinal plants for analysis in the US?

Obviously, the ownership of ideas (whether scientific inventions or traditional knowledge) has been the source of far more wealth than the ownership of genetic resources in their physical state. The global trade in plant genetic resources is big business, estimated by ten Kate and Laird (1999) to be several billion dollars annually, so it's hardly surprising that the scramble to get legal protection for products and processes has had some significant impacts on developments in intellectual property law – developments that have been almost entirely shaped by plant genetic resources issues.

As we mentioned in the previous chapter, the number of genetically modified fish strains developed in the past couple of decades is infinitesimal compared to the number of plant strains. Pharmaceutical companies are struggling to find powerful cures based on marine organisms, but the vast majority of drugs inspired by compounds existing in nature are based on plants – by 1996, only two of the top selling 150 pharmaceuticals were derived from marine organisms (Brush and Stabinsky, 1996). Nevertheless, there may be sufficient parallels between some uses of plant and aquatic genetic resources – for example, between plant and marine bioprospecting and between farmed fish and crop enhancement – that patent law agreements can safely accommodate both types of activity.

Where the intellectual property comparisons between the two fields diverge most dramatically is in the area of traditional knowledge. There is simply no parallel in traditional fishing villages to traditional farmers' knowledge of crop breeding and to other communities' knowledge of medicinal plants. This might seem a matter of no great note were it not for the fact that several emerging national biodiversity laws, based loosely on the CBD guidelines, spell out the rights of traditional communities in terms of the knowledge they contribute to the use of aquatic genetic resources. There is no doubt that traditional fishing communities have a wealth of knowledge that comes from the importance of fish as food or ceremonial object – including knowledge of fish life cycles, migration habits, habitat preferences and of techniques for catching fish – yet it is likely to be largely irrelevant to uses for breeding. As we'll show in this and the following chapters, a too rigid interpretation of laws tying community rights to traditional knowledge could mean significant future conflicts (as the trade in aquatic genetic resources becomes more common) and run counter to the CBD's objectives of promoting conservation, sustainable use and equitable sharing of benefits.

Fighting for the Right to Own Genes and Genetic Inventions

By protecting the ideas of inventors from unauthorized use by others, intellectual property law (including patents, copyright, trademarks, industrial designs and trade secrets) protects an inventor's economic interests and the interest of society in encouraging new inventions that may benefit the public. A patent is a legal certificate that gives an inventor exclusive rights to produce, use, sell or import an invention for a fixed period, usually 17 to 20 years. A patent application must demonstrate that an invention is useful (have industrial application), novel (recent, original and not already publicly known), not obvious to a person skilled in the technology and more inventive than mere discovery of what already exists in nature. Patents can be granted for products, specific uses of products (eg use of a drug to cure cancer but not for purposes yet to be discovered), and processes used to create a product.

The law of patents was developed in 19th-century Europe to protect inventions of factory machinery and excluded the protection of living materials, foods and medicines. However, demands for protection of inventions related to commercial uses of genetic resources have revolutionized patent criteria, and the state of the law remains in flux in the wake of new scientific advances and amid controversies about the patenting of life and the role of traditional knowledge in invention. In the 1960s, the US legislated the right of plant breeders to patent seeds, preventing others from selling the same variety, and in 1980 the US Supreme Court accepted the first patent on a genetically altered micro-organism. By 1997, the US had granted 69 patents on animals (Correa, 1999).

Keeping track of the current state of the law is further complicated by the fact that different countries take different approaches to patent law. The European Patent Convention provides that microbiological processes are patentable but 'essentially biological processes' are not, thereby excluding plant varieties obtained by conventional breeding. No such restriction exists in the US (Crucible Group, 1994). Decisions by the US Patent and Trademark Office to grant monopoly rights over plant, animal and human genetic materials have sparked a rush to collect, map and patent genes, based largely on their potential.

Developing countries have been under pressure to recognize US patents, although many developing countries exclude patent protection for plant varieties and animal races (Correa, 1999). Both the scope of protection and the rights of patent holders continue to be expanded as the law continues to evolve in industrial countries. By contrast, developing countries such as Argentina, Brazil and India have set limits by allowing patents on processes but not products and by requiring patent holders to make socially useful products available in the domestic market (Grenier, 1998).

Genetic modification of animal life (the 'Harvard mouse' is a controversial example) has drawn attention to the need to address demands to allow patents on living creatures. Pharmaceutical companies and bioprospectors collecting on their behalf have been actively patenting processes to develop drugs and cosmetics derived from research based on aquatic genetic resources that range from marine invertebrates to algae. Patenting of new fish varieties (such as the Super Salmon and Arctic char varieties) and of processes used in their development has been more limited but can be expected to increase significantly as the aquaculture industry expands. The US Patent and Trademark Office led the way by approving a patent protecting a method of increasing the growth rate of a transgenic salmon (Correa, 1999).

The adaptation of patent laws to cover genetic resources is still controversial. A cotton gin is a very different type of invention from a cotton (or catfish) gene. Patenting of living materials not only raises ethical questions, it also frequently raises concerns about the novelty of 'inventions', especially when an inventor may have relied for inspiration on traditional knowledge. Historically, while technology exporting countries have been quick to develop patent legislation to promote new development, technology importers have had little reason to develop their own patent laws. Developing countries have been understandably reluctant to embrace universal patent laws relating to uses of genetic resources (as TRIPS proposes) because applicants from industrialized countries are the primary beneficiaries (GRAIN, 1998a). As Halewood (1999) notes, the net effect of globalization of intellectual property laws through agreements such as TRIPS and the North American Free Trade Agreement has been primarily to benefit developed countries in the use of genetic resources.

Plant breeders have pushed for an extension of intellectual property rights to include new varieties produced by breeders. There are no equivalent mechanisms for aquatic genetic resources, but these are likely to develop, albeit with modifications that reflect the fact that few aquatic animals have been domesticated (Bartley and Pullin, 1999).

TRIPS: Controlling Access to Genetic Inventions

All countries that are members of the World Trade Organization (WTO) are obliged to implement TRIPS. The agreement focuses on the importance of intellectual property rights in promoting trade liberalization. The emphasis on intellectual property in WTO agreements is a response to the explosive growth in information technology and biotechnology in international trade as well as the desire of some industrial countries to protect products from intellectual 'piracy' in foreign markets (Crucible Group, 2000).

TRIPS requires all member states to set minimum standards of intellectual property protection, thereby ensuring a far higher degree of global uniformity than previously existed. Patents must be available for inventions in all fields of technology without discrimination, whether products are imported or locally produced. Article 27 requires member states to adopt national level intellectual property systems for all products and processes, including pharmaceuticals, modified micro-organisms and microbiological processes. Unlike the earlier international undertaking, TRIPS clearly applies to inventions derived from aquatic genetic resources.

Negotiations on the passage of TRIPS highlighted the very different viewpoints of industrial and developing countries on the extent to which patenting of biological diversity-related inventions should be permitted. The most controversial section of TRIPS has been Article 27.3(b). While it allows countries to exclude plants and animals as well as essentially biological processes from patentability, WTO members must provide protection for plant varieties either by patents and/or by an effective *sui generis* system – that is, a unique system of rights for a specific item or technology. The exemption for plants and animals remains a controversial issue, with some developed countries facing strong pressure from industry to push for the removal of the exemption. Some critics of TRIPS argue that doing so would undermine the CBD by depriving countries of the right to prohibit IPRs on life forms and by diminishing their ability to negotiate a fair share of benefits arising from the use of genetic resources (GRAIN, 1998a). During a 1999 review of

TRIPS, a group of least-developed countries proposed, unsuccessfully, that TRIPS should contain a provision that patents must not be granted without the prior informed consent of the country where the genetic resources originated (Martinez, 2002). In 1998, an Organization of African Unity (OAU) model law proposed a prohibition against patents on inventions derived from biological resources obtained from member countries.

In essence, developing countries were pressured into accepting TRIPS in return for the promise of the economic advantage created by improved opportunities for trade. Some developing countries objected to the imposition of Western notions of 'inventions' in parts of the world where many important innovations had occurred in an evolutionary and informal manner in farming and other local communities.

Indigenous groups have condemned TRIPS as flying in the face of indigenous traditions of treating community knowledge as a shared resource and as essentially forcing indigenous peoples to adopt Western patent laws in order to prevent the commercial appropriation of their own knowledge. In the opinion of one group, TRIPS pushes indigenous peoples to 'play in a game in which the rules are defined by the opponents' (Tauli-Corpuz, 1999).

WHO OWNS TRADITIONAL KNOWLEDGE?

What happens if, without a community's permission, a company patents a product or process made possible by the use of the community's traditional knowledge?

While it is true that many indigenous communities have a history of freely sharing their knowledge about wild plants and animals, they may have concerns about the unauthorized use of their knowledge that have nothing to do with compensation but a lot to do with issues such as a lack of respect for a living creature or deceitful dealings with a community, etc. Patent offices rarely require applicants to disclose whether an idea has been appropriated from a traditional community. Even if a traditional community wanted to seek recourse, using the avenues of Western legal systems might be both contrary to its beliefs and unaffordable.

Interest in access to traditional knowledge about plant genetic resources has grown by leaps and bounds as the gene rush seeks out new sources of potentially valuable information – like the gold rushes of earlier centuries, making a valuable find may be a long shot, but the potential rewards may be so high that it's worth the effort and the risk. That's why the question of who owns traditional knowledge and how communities can control its use has become such a big issue in recent years, as illustrated in the section below on biopiracy. It's also why international agreements like the CBD have placed so much emphasis on using the demand for traditional knowledge as a rationale for the expansion of community rights. As we'll see in the next chapter, while this may be a sensible approach in the plant world, it may have little relevance for aquatic genetic resources.

The Wapishana Go to Court: The Case of the Fish Killing Plant

The Wapishana are experiencing what many other indigenous groups and local societies are suffering with alarming frequency throughout the tropics: the appropriation of traditional knowledge by pirates passing for scientists, researchers, missionaries, environmentalists, activists of indigenous people's rights, and other disguises.

JULIO CESAR CENTENO (CENTENO, 2000)

Several Wapishana indigenous communities living in the Amazon basin along the Brazil-Guyana border have long used their knowledge of the toxic properties of the cunani bush to catch fish. When the chewed leaves are thrown into a river, fish in the immediate vicinity reportedly leap out of the water and soon die. The fish can be eaten immediately, with no alteration in their taste and no side effects on humans.

A British biochemist who had spent years living with the Wapishana and studying their traditional uses of plants heard about the unusual properties of the cunani. After he left the country, he undertook extensive research to isolate active ingredients of some of the plants he had collected. He then registered patents in the US and Europe on active ingredients in cunani that are believed to act as powerful stimulants to the nervous system or as neuromuscular agents that can prevent heart blockages. He also patented active ingredients of tipir, the nut of the greenheart tree, which the Wapishana had long used to stop bleeding and prevent infection and is being investigated for anti-malarial properties.

When Wapishana chiefs heard what the biochemist had done, they accused him of stealing the knowledge of their ancestors and elders in order to sell it to pharmaceutical companies. As one Wapishana woman put it, 'This knowledge has always been with the Wapishana. It's part of our heritage and now is being taken from us without any payment'.

The biochemist expressed puzzlement at the uproar. He pointed out that his discoveries were the result of a lifetime spent decoding the ingredients in traditional Wapishana remedies. The patents were justified, he argued, by the results of his own intellectual effort (Singh, 2000a; Veash, 2000). After an extensive campaign, the Wapishana people eventually succeeded in overturning the cunani patent but not the tipir patent. In addition, at least one chief prohibited all future visits by researchers, whatever their purpose might be.

Drawbacks of IPRs Protection of Traditional Knowledge

The TRIPS agreement sparked considerable protest among advocates of indigenous and local communities who believed that it would erode the already fragile rights of communities to control the use of their knowledge. One of the biggest issues has been whether a product derived from traditional knowledge can truly be called an invention – one of the prerequisites for the right to patent.

After the TRIPS agreement, the World Intellectual Property Organization (WIPO) attempted to address the issue through its Intergovernmental Committee on Intellectual Property and Genetic Resources. The Committee looked at ways to take traditional knowledge systems into account in the development of IPRs systems through *sui generis* approaches – that is, unique mechanisms for providing legal protection for holders of traditional knowledge.

Some indigenous movements have been sharply critical of the WIPO approach, arguing that *sui generis* protection of traditional knowledge rights is nothing more than an attempt to assimilate indigenous rights into 'Western' property systems and bypass the real issue – the rights of indigenous peoples to direct control over lands and resources in traditional territories.

Opinions are divided about the usefulness of IPRs in protecting community interests. Some people believe that IPRs as they currently exist can serve the purpose by allowing users of genetic resources to generate revenues from their inventions and share the revenues with countries of origin or local communities. Others believe that existing IPRs are not only inadequate to protect the rights of indigenous peoples but also represent one of the greatest threats to the conservation of biodiversity (Crucible Group, 2000). Indigenous and local peoples generally express a desire to link the protection of indigenous and local knowledge with territorial rights and the right to self-determination – a position that has been reiterated in a wide variety of international and local peoples' declarations and in statements in international multilateral fora (Halewood, 1999).

Theoretically, IPRs could be used to protect the knowledge and innovations of indigenous and local communities. However, differences in Western and indigenous legal concepts make it impractical. Communal property is the prevailing system used in most traditional societies to control access to basic resources, and even in cases where esoteric knowledge is the exclusive intellectual property of individuals, families, shamans, clans or lineages, these owners cannot necessarily commercialize the knowledge without the permission of the whole community or tribal elders (Posey and Dutfield, 1996).

The use of patent law as it currently exists is also problematical because of the difficulty of defining the source of traditional knowledge, accumulated over generations, and therefore its novelty. Use of customary laws is another option, but many nations do not recognize customary law and national legal systems may conflict with unwritten customary laws. Finally, indigenous peoples view any use of intellectual property law with deep suspicion because of its association with past exploitation of traditional knowledge. Shiva (1997) characterizes intellectual property law as a modern form of colonization extended to the interior spaces, the 'genetic codes' of life forms, made possible by the treatment of genetic resources as a common heritage – a view that resonates with many indigenous rights groups.

Sui Generis Protection of Community Rights

The inadequacy of patent law as a means of protecting traditional knowledge has led some community rights advocates to propose the use of *sui generis* alternatives to achieve the same end. '*Sui generis*' refers to rights that are designed to be unique for a specific purpose and are not covered by existing legal systems. The TRIPS agreement requires WTO member countries to protect plant varieties either by patents or an effective *sui generis* system, but does not define what that means.

The term has become common jargon without any universal understanding of its definition, perhaps in large part because *sui generis* systems are largely still in the formative stages. Canadian land claims agreements with indigenous peoples, referred to in Case Studies 2 and 3, are one recent example of *sui generis* recognition; recent laws in countries like the Philippines that are specific deal to the protection of community rights are another.

One proposed model for *sui generis* protection is a system of community intellectual property rights that would establish the legal right of communities to protect knowledge developed communally rather than by an identifiable individual – primarily to protect farmers' rights to control access to seed. The Crucible Group (1994) suggested that, in order to be effective, community intellectual property rights need to be entrenched in national legislation with reciprocal recognition by other countries, supported by an international database for tracing germplasm. Under such a system, plant varieties developed by communities might be deposited in germplasm banks along with registration data (such as the date, place and community of origin), enabling the source to be determined if complaints arose about appropriation of traditional knowledge by outside interests.

While there are no analogous issues in the use of fish germplasm, a community IPRs system could conceivably be useful for the protection of other types of traditional knowledge of aquatic genetic resources – for example, by providing for the documentation of knowledge of medicinal uses. One model community intellectual rights act, suggested by a Third World Network discussion paper, proposed meeting the patent law novelty requirement by describing indigenous peoples as innovators because they have developed knowledge unknown to the outside world (Singh Nijar, 1998). Posey and Dutfield (1996) suggest that such a law would be compatible with the TRIPS call for *sui generis* IPRs protection and with Article 8(j) requirements. As discussed in the following chapter, this type of legal protection has far more relevance for terrestrial than for fishing communities because examples of traditional innovations using aquatic genetic resources are far less frequent.

Posey and Dutfield (1996) propose a more expansive system of 'traditional resource rights' that moves beyond the narrow limitations of knowledge protection. In their view, existing international agreements (eg on human rights, right to self-determination, land and territorial rights, and intellectual and cultural property rights) provide the basis for overlapping and mutually supportive 'bundles of rights' that would encompass not only IPRs protection but also the control of resources, which indigenous peoples view as central to self-determination. Posey points out that while the privatization or commoditizing of property is often contradictory to indigenous peoples' spiritual beliefs, indigenous and traditional communities are increasingly involved in market economies, and that the right to development and conservation are mutually supportive. Traditional resource rights would also recognize the inextricable link between cultural and biological diversity. Posey suggests that they could be implemented locally, nationally or internationally and could guide both international law and national legislation.

BIOPIRACY: PLAIN DEALING OR PATENT THEFT?

The urgency of addressing the question of community rights over genetic resources has been heightened in recent years by a flurry of complaints about biopiracy. The Crucible Group

(2000) defines 'biopiracy' as the taking of genetic or biological resources without the prior informed consent of local people or a competent state authority for access and benefit sharing under mutually agreed terms.

Most frequently, charges of biopiracy are related to the use of plant genetic resources and often are spurred by patents of discoveries that are in some way related to the use of traditional knowledge of farmers or of indigenous communities. One much publicized example was the attempt by a US company to patent Basmati rice. Another was the dispute between the pharmaceutical company researcher and the Wapishana people.

Had there been an access and benefit-sharing law in place, the Wapishana might have negotiated an agreement with the biochemist to share in the profits from any future royalties or obtain some other more immediate reward, or they might simply have denied him permission to make collections or refused to give him any information about their traditional uses. They might also have demanded benefits that he felt unable to provide without knowing whether his research might lead to a deal with a pharmaceutical company or the development of a marketable product. Article 8(j) of the CBD is intended to deal with exactly this type of situation. Situations like the one involving the Wapishana have occurred many times in different parts of the world, as traditional communities attempt to earn recognition for the value of their plant knowledge, and examples of benefit-sharing agreements are beginning to occur with more and more frequency.

Biopiracy and Aquatic Genetic Resources

It's hard to imagine aquatic resources parallels for the biopiracy examples described above. One might have occurred if Aqua Bounty had developed and patented the 'Super Salmon' after learning from indigenous people about the anti-freeze properties of the ocean pout, which were used to create the new strain. Even then, the company might have argued that the desirable characteristics of the pout were obvious to any observer. Traditional knowledge of aquatic biodiversity may be abundant but is not necessarily relevant to uses of aquatic genetic resources such as aquaculture and the development of pharmaceutical products from animals that live in areas that are inaccessible to local people.

Nevertheless, complaints about biopiracy of marine organisms are not uncommon. Indeed, one of the main catalysts for the passage of access legislation in the Philippines in 1995 (see Case Study 5) was a series of newspaper articles on biopiracy of marine invertebrates (and of plants) by foreign collectors. After the law came into effect, the complaints continued, and communities began targeting the government for not properly enforcing the law and for its lax requirements for obtaining community consent.

When the US-based pharmaceutical company Neurex Inc isolated and patented the toxin SNX-111, produced by the Philippine sea snail *Conus magnus*, civil society organizations took both the company and government to task. The stakes were large: sales of the pain-killing drug derived from the toxin reportedly earned Neurex more than US$80 million during the first year of marketing. Critics charged that the government supported biopiracy by funding research by scientists at the Marine Science Institute (MSI) who collaborated with the University of Utah in collecting specimens, isolating the toxin and forming a private company to capitalize on the snail (Bengwayan, 2001).

Previously, collectors for the MSI would purchase specimens from fishermen in several areas of the country, who gather different varieties of cone snail for their ornamental value and have no use for the meat.[11] Philippines law (Executive Order 247) now requires foreign collectors to partner with Philippines research institutions, which must get community consent for collections (whether or not traditional knowledge is needed by collectors). To comply with regulations under the law, MSI holds public information sessions in communities where collections of marine organisms are proposed and must also obtain permission from local government representatives. Critics of the process note low turnouts at information sessions and suggest that local people may not even be aware that marine bioprospecting is taking place in their communities (Batungbacal, 2000).

Making informed consent requirements both fair and feasible is a tough challenge. If the consent process is too cumbersome and expensive, collectors may simply go to other, more 'friendly' countries if the organisms they're looking for can be obtained elsewhere. For both countries and communities, it's a question of how to develop effective regulations without killing the goose that might lay a golden egg.

The Philippines law distinguishes between collections for academic and commercial purposes. Getting a fair share of whatever profits may arise from research by foreign collectors continues to be an issue for countries that have not yet passed laws regulating access. In the Bahamas, University of California researchers obtained a permit for the collection of soft coral, without charge and with only the condition that they file reports on their work. The university subsequently obtained a patent on a production process that was in turn sold to a company that marketed it to the cosmetic company Estée Lauder, which used it in the development of beauty products. The Bahamian government unsuccessfully attempted to negotiate royalty payments from the extracting company, which responded that it might simply move its collection activities to more accommodating Caribbean countries where the same species is abundant. The Bahamian government is now in the process of developing legislation to govern access to genetic resources.

University of California researchers have also been criticized for patenting the anti-inflammatory agent pseudopterosin, a compound found in sea whips in the Caribbean and subsequently incorporated in a skin cream also marketed by Estée Lauder. The average annual royalty income received by the university for patented pseudopterosins is reported to be more than US$750,000 (Singh, 2000b).

While bioprospecting for marine invertebrates gets most of the attention, biopiracy concerns sometimes extend to fish as well. The Amazon region has been hard hit by illegal trade in wildlife, and the Brazilian government attempts to be vigilant in stemming its flow. However, given the size and remoteness of the region, enforcement of environmental law is a stupendous task. In another example of suspected aquatic biopiracy, a

German aquarist who planned to breed ornamental cichlid varieties (cichlids include important food fish such as tilapias and tiny ornamentals prized for their bright colours) travelled to Brazil to collect wild specimens in the middle Rio Negro. When apprehended by a representative of IBAMA, the Brazilian Institute responsible for enforcement of protected species laws, the collector was examining the viscera of cichlids to determine their natural diet. He had apparently obtained written permission from the nearest municipality but had not sought the necessary approvals from IBAMA. He was deported and ordered not to return.

Biopiracy Debates in Brazil

The chief prosecutor for the Brazilian State of Amazonas, who opened an inquiry into biopiracy in 1997, estimates that about 20,000 individual plant samples are illegally removed from the country every year. Scientific laboratories generally receive the samples and information from third parties who don't say where they came from.

How is the information obtained? A representative from IBAMA notes: 'The scientists congregate in small frontier towns. Then they ask the Indians what they would do if they had a headache, muscle pains or a bad stomach. The local people then take them into the jungle and show them which plant they would use to cure those symptoms. The scientists pay the Indians a little money, then take the plant back to their labs. There, they discover the principle by which the plant works and sell their preliminary research on to the pharmaceutical companies for development'.

Robert Smeraldi, director of the Friends of the Earth Amazon programme, notes: 'There is widespread smuggling of genetic material by unauthorized companies. I'm not talking about respectable pharmaceutical companies being directly involved, but I do believe they could benefit from illegal research'.

A Brazilian cancer specialist was sharply criticized by the media after receiving close to US$1 million from the American National Cancer Institute to analyse a number of Amazonian plants. His critics argued that if information taken from indigenous peoples is used to develop the world's first anti-cancer drug, the Indian tribes might lose out even if the cancer specialist himself does not profit.

In his defence, the researcher noted: 'While we continue to talk and worry about biopiracy, fewer people are out there actually studying the Amazon, which is a serious form of scientific neglect. Of course biopiracy happens, but we have to balance this against not researching the rainforest at all. Ultimately that is much more damaging to mankind' (Veash, 2000).

The money that changes hands in the collection of ornamental fish can stymie the most determined efforts at enforcement, especially in areas like the Amazon – the largest and perhaps most ungovernable river basin in the world. Rohter (2001) reports that zebra fish from the Xingu River are so valuable that specimens are used as local currency and that dealers' profit margins are comparable to those in the cocaine trade, with a particularly rare specimen bringing as much as US$600 from a Japanese collector.

Although Brazil has attempted to control sales to prevent rare species from being wiped out, black market dealers continue to thrive. Rohter quotes a regional IBAMA representative as stating that, with only five agents to monitor all wildlife in an area twice the size of New Jersey, little can be done to protect illegal traffic. A tropical fish expert at the Emilio Goeldi Museum in Belem notes that tropical fish dealers routinely file false customs declarations and shipping waybills to get around restrictions, yet that there is little public interest in the problem (Rohter, 2001).

Stopping unauthorized collection of aquatic genetic resources is a challenge at the best of times. Access laws are unlikely to work unless government has the political will and provides the necessary funds for enforcement. This support may be absent in both developed and developing countries depending on the degree of cooperation among different levels of bureaucracy and of coordination among policy structures. Bribery of officials may create an additional impediment in countries where it is a common tradition.

Opposing Views on Biopiracy

Even when the approval of national agencies and indigenous communities is received, charges of biopiracy may still arise. Indigenous communities may grant permission for collections to occur without understanding (or in some cases being adequately informed of) the true purpose of the collection, as in the case of the Wapishana and the cunani plant. Collectors may take unfair advantage of communities that lack the capacity or the necessary information to negotiate fair deals. Even if the purpose of collection is fully understood, agreements may be perceived to be grossly unfair. Indigenous people may receive a pittance for the right to collect or may be offered royalties that never come to pass or are minuscule in relation to profits. A 1998 estimate placed annual global sales of pharmaceuticals at US$300 billion, of which between 25 and 50 per cent are derived from genetic resources (ten Kate and Laird, 1999). With almost 70 per cent of the total number of ethnolinguistic groups in the world living in 225 regions of the highest biological importance (WWF, 2000), it may be assumed that indigenous communities have been and will continue to be primary sources of the genetic resources needed for pharmaceutical development.

As the Wapishana example illustrates, biopiracy applies both to physical resources and the knowledge that makes their value immediately evident to collectors. Communities face challenges on both fronts: protecting their knowledge against misuse or appropriation under foreign intellectual property laws; and protecting resources over which they may have no legal authority. Very often, in the eyes of communities, the greatest perpetrators of biopiracy are not foreign interests but levels of government in their own countries that exercise authority over 'public' resources nationwide and deny indigenous and other communities a say in who collects plants, fish and animals in community territories. In several countries, struggles for the right to self-determination are fuelled in large part by frustration over lack of access to traditional resources and exploitation by outside interests who have no obligation to seek community consent once having received the consent of governments. Corporations may face accusations of biopiracy even when they have conscientiously complied with all government regulations or have negotiated with communities in what they believe to be good faith. So may collectors from scientific institutions who depend on access to genetic resources for basic research, and whose findings are freely published.

Accusations of biopiracy may relate to activities ranging from the collection of individual animals (eg rare Brazilian ornamental fish) to the entire system of collection of biological and genetic resources – what Shiva (1997) describes as the sequel to colonization of countries, in which 'colonies have now been extended to the interior spaces, the 'genetic codes' of life forms from microbes and plants and animals, including humans'. As Dutfield (n.d.) points out, noting that exploitation of genetic resources has even been called 'the slavery of the new millennium', such rhetoric may in fact undermine legitimate efforts to gain fairer treatment for countries and communities that seek a more level playing field. While the term 'biopiracy' may sound inflammatory to some, the issue raises legitimate questions about fair play for countries and communities that provide aquatic genetic resources.

AN INDIGENOUS COMMUNITY SAYS NO

Countries take many approaches to the recognition of indigenous rights of ownership and control over aquatic genetic resources. Canada has been actively involved in the negotiation of treaties with indigenous peoples. The Inuit, a people along the Arctic coast, recently completed an agreement that recognizes both land and resource rights. The Canadian DFO retains a role in the management of sea-going fish, but communities have the right to prior informed consent to collection of fish broodstock.

Arctic charr, the northernmost species in the family Salmonidae, has long been a staple in the diet of Inuit peoples on the coast of the Arctic Ocean in Canada. While some charr inhabit landlocked lakes, most are anadromous, migrating in late summer from the ocean to rivers and lakes to spawn, and reaching a weight of up to 11.3 kg. During recent years the popularity of the fish in urban markets has increased, creating an additional source of income for Inuit fishers in northern Canada. But charr are slow growing, and the genetically unique subpopulations or stocks of charr are extremely sensitive to overfishing. Consequently, Inuit communities developing sustainable management plans have had to grapple with how best to conserve a diminishing resource while supporting local fishing economies (eg Holman Hunters and Trappers et al, 1994).

The Northwest Territories Scientists' Act requires consent of local communities before any scientific research is undertaken in the Territory. Enacted in 1974, the Act was one of the earliest examples in a developed country of legislation that reflects emerging international principles for respecting indigenous knowledge and returning benefits to knowledge holders and their communities (Mann, 1997). In 1993, Canadian Inuit concluded negotiations of a land claim agreement with the government of Canada. The Nunavut Land Claims Agreement established principles of Inuit priority in the harvesting of marine resources and of ownership over the resources in Inuit-owned land and marine areas.

Soon after the 1993 land agreement came into effect, Canada's DFO began refusing fish harvesting, research and farming permits in the Nunavut region unless the prior consent of local communities had been obtained. Icy Waters, the major charr farming company in the Arctic, proposed a joint venture with Inuit communities and an Ontario university research group to set up a new company, Suvaaq Inc., to improve the company's existing broodstock that were based on previous DFO collections. Under the proposal, each of seven participating Inuit communities would receive a 5 per cent equity stake in the new company in exchange for sperm from six male Arctic charr from two separate stocks found in waters near the communities. Icy Waters suggested that Inuit communities would also benefit through education and practical experience in fish farming and access to genetically improved stocks as these were developed. The business proposal provided that each community would own its original fish contribution but that hybrid lines resulting from cross breeding would be owned by Suvaaq (Mann, 1997). The proposed project would result in Icy Waters gaining access to a total of 14 genetically distinct charr stocks through local communities.

Final approval of the proposed project required consent from the communities (through local Hunters and Trappers Associations), DFO and the Nunavut Wildlife Management Board (Mann, 1997). The latter body, established under the land claims agreement to oversee the protection and wise use of wildlife and wildlife habitat for the benefit of the Inuit and other residents, advises the governments of Nunavut and Canada, which maintain decision-making authority. The nine-member board includes four Inuit representatives and four from the Canadian and Nunavut governments.

Several difficult issues emerged during negotiations on the proposal. Local fishermen worried that the sale of genetically improved farmed fish would have a negative effect on markets and prices for wild caught fish (Nunavut Wildlife Management Board, 1998). Icy Waters attempted to allay this fear by suggesting that successful farming could benefit local fisheries by reducing commercial harvesting pressure on wild stocks, ensuring a valuable sport fishery, and increasing consumer awareness of charr.[13] Ownership issues added a further complication. These included not only concerns about Suvaaq ownership of successive generations of charr hybrids but also the possibility that the university research group might try to obtain a process patent based on genetic mapping of charr (Mann, 1997). Mapping would accompany the collection of genetic resources and was necessary in order to ascertain whether the genetic differences between the collected populations were significant. Finally, some Inuit expressed concern that the project showed a lack of respect for charr and that the spirit of the charr might take revenge on the Inuit people if the project went ahead, a not unusual sentiment among native people who may accept technologies as a necessary evil yet still feel uncomfortable with the spiritual implications of altering nature.

Ultimately, communities withheld their consent and the Wildlife Management Board turned down the proposal. Icy

Waters was later able to obtain broodstock from two charr stocks from one community that had consented to their collection prior to the land claims agreement, and the company still pays royalties to that community. These broodstock were subsequently cross-bred with the company's base stock.

Icy Waters described the failed negotiations as expensive and time-consuming, complicated by the difficulty of dealing with several different levels of authority, the need to negotiate with several communities over a vast land area, a high level of confusion about the implications of fish farming, and a long history of local suspicion of southern interests.

Some of the issues that characterized the Icy Waters negotiations – such as the patenting of products and processes – surface frequently in negotiations of agreements for access to genetic resources. However, the Icy Waters example also illustrates that collectors of aquatic genetic resources may face extra complications that are unlikely to occur in negotiations for access to plant genetic resources. For example, the need for genetic diversity of broodstock may necessitate negotiations with several communities over a wide area (because genetic differences, at least in salmonids, usually reflect geographic separation), and in different government jurisdictions, adding to the cost and complexity of negotiations.

Indigenous communities may also hold spiritual beliefs that create significant concerns about the movement of fish from their natural habitat – in addition to potential environmental concerns about introduction of fish into areas where they are not indigenous. Suspicion or confusion about possible effects of fish farming on local fishing economies is common, and will also hamper negotiations. And because government policies for fish conservation and use are still at the stage where the genetic fine structure of fish populations is poorly represented (and, in fairness, usually not even known), the ground-rules are often less clear than in the plant world. The Icy Waters case is typical in that genetic mapping of the collected stocks would have taken place as they were collected. In other words, the genetic distinctness of the resources in question was not even known.

What chiefly distinguishes the communities involved in the Icy Waters negotiations from many others in the world is that they have substantial degree of control over the of use their genetic resources and the clearly recognized authority to withhold consent.

Most indigenous and local communities around the world (and in Canada, for that matter) have no such authority. This is one among many factors complicating the efforts of countries to develop access laws and policies that acknowledge in a meaningful way the importance of involving local and indigenous communities in decisions regarding applications for access.

Fish and shellfish are nutritious foods that constitute desirable components of a healthy diet. Most seafoods available to the U.S. public are wholesome and unlikely to cause illness in the consumer. Nevertheless, there are areas of risk. The major risk of acute disease is associated with the consumption of raw shellfish, particularly bivalve molluscs.

For persons living in areas in which reef fish are consumed (Hawaii, Puerto Rico, the Virgin Islands), there is a risk of ciguatera; other natural toxins (paralytic shellfish poisoning, neurotoxic shellfish poisoning, etc.) have been associated with shellfish from endemic areas. Finally, there are less well-defined risks of acute and chronic disease related to environmental contamination of aquatic food animals. Dealing with such risks on a short-term basis requires improvements in the present system of regulatory control. In the long term, amelioration and eventual elimination of some hazards require strengthening and more effective application of control measures to prevent the disposal of human and industrial waste into offshore marine and fresh waters.

Because of the strong public interest in seafood safety and the declared intention at the congressional level to develop a new inspection system, a clear opportunity exists to introduce innovative methodologies for control that address directly the important health issues associated with seafood consumption.

This report reviews the nature and extent of public health risks associated with seafood, and examines the scope and adequacy of current seafood safety programs. The conclusions and recommendations arrived at are summarized in the following material:

- Most current health risks associated with seafood safety originate in the environment and should be dealt with by control of harvest or at the point of capture. With minor exceptions, risks cannot be identified by an organoleptic inspection system.
- Inspection at the processing level is important to maintain safety of seafoods, but there is little evidence that increased inspection activities at this level would effectively reduce the incidence of seafood-borne disease.

With currently available data, it is possible to identify the source of much of the acute illness associated with seafood consumption, though the dimensions of the problems are not always known; these data, in turn, can form the basis for national control programs.

- Chronic illness resulting from seafood consumption is associated primarily with environmental contamination; thus, control depends on improved understanding of the occurrence and distribution of the chemical agents involved, the exclusion of contaminated seafood from the market, and increased action to prevent additional pollution of the waters.
- Because well over half the nation's seafood supply is imported and environmental contamination is globally pervasive, it is important that the safety of imported seafood be ensured through equivalent control measures in exporting countries.
- One-fifth of the fish and shellfish eaten in the United States is derived from recreational or subsistence fishing, and these products are not subject to health-based control; there is need to improve protection for consumers of these products by regulation of harvest and by education concerning risks associated with their consumption.
- Because the problems are largely regional, the primary effective control – except for imports – is at the state

level, and this effort should be strengthened. However, there is need for federal oversight, general rule setting, and support to ensure the effectiveness of state-based programs and to provide expert assistance and specialized facilities.

- There is a lack of understanding of the nature of seafood hazards in the food service sectors and by the consuming public and health professionals; a vigorous campaign for information dissemination and education in these matters is needed, particularly for high-risk consumers and high-risk products such as raw shellfish.
- An improved national surveillance system should be developed to provide more reliable and comprehensive information on seafood-borne disease incidence. Data will then permit meaningful risk identification and risk assessment as a basis for effective regulation of seafoods (current data on disease occurrence in seafood consumption are too fragmentary to allow reliable risk assessment of microbiological and natural toxin hazards).

A summary of hazards, risks, and their control for the major groups of hazardous seafoods in shown in Table 1-1. They are arranged in order of importance.

Among seafood consumers, the group at greatest risk appears to be consumers of raw molluscs because of environmental contamination and naturally occurring vibrios. Consumers of recreational and subsistence fishery products are the second largest constituency at risk, both from natural toxins and from environmental contaminants. Inhabitants of islands in the Caribbean and tropical Pacific Ocean, and to a lesser extent of the Gulf of Mexico States, are uniquely at risk from ciguatera due to consumption of fish that feed on or near tropical reefs and prey upon reef fish. Although U.S. mainland inhabitants have developed the disease from imported fish, ciguatera is a highly regional problem that greatly skews disease reporting because of its geographical concentration and its high visibility in affected regions. There is a more general risk to consumers from certain fresh and frozen fish due to a milder toxic disease, scombroid poisoning.

There is generally a low risk to seafood consumers of food-borne disease due to control failures during processing or at the food service level. Fish do not normally carry microorganisms such as *Salmonella* or *Campylobacter*, which are commonly found in land animal carcasses and are the major causes of reported food-borne disease (though fish may acquire bacteria from contaminated water). Some ethnic practices in the preparation of fish for eating, place a small number of people at high risk from botulism, but this is not a significant hazard for most consumers of fish.

Thus, the health risks associated with seafood – although diverse – are identifiable and, to a significant extent, controllable by innovative measures aimed at geographically restricted or species-specific problems. Some risks, particularly those associated with environmental contamination, may be increasing; their elimination will require a major commitment on the part of both government and industry to change methods of waste disposal in our society. These and the more visible hazards mentioned can be greatly mitigated by a regulatory system specifically aimed at the causes, be they natural toxins, microorganisms, or contaminants. However, this will require something other than organoleptically based inspection systems, which may be useful for quality control and grading but are essentially worthless for detecting and controlling health risks.

5

Naturally Occurring Fish and Shellfish Poisons

Incidents of illness due to naturally occurring seafood toxins reported to the Centers for Disease Control in the period 1978-1987 were limited to ciguatera, scombroid fish poisoning, and paralytic shellfish poisoning. Other intoxications, including puffer-fish poisoning and neurotoxic (brevetoxic) shellfish poisoning, were reported earlier, and diarrhetic shellfish poisoning and amnesic shellfish poisoning are prospective risks that should be anticipated.

Naturally, toxic fish and shellfish cannot be distinguished from nontoxic animals by sensory inspection, and the toxins are not destroyed by normal cooking or processing. Except for scombroid fish poisoning, natural intoxications are both highly regional and species associated, and toxins are present in the fish or shellfish at the time of capture. Scombroid poisoning is due to histamine produced by bacteria multiplying on certain fish that are mishandled after capture, and illnesses are widely reported from different states. Ciguatera is a sometimes severe disease caused by consuming certain species of fish in tropical waters usually associated with islands or reefs.

The disease is most common (endemic) in the Caribbean and Pacific islands, with some outbreaks in southern Florida and sporadic cases in other states due to imported fish or tourist travel to endemic areas. Ciguatera was responsible for about half of all reported outbreaks of seafood intoxications in 1978-1987. Treatments are largely supportive, but mortality is low. There are presently no effective control systems in place for prevention of ciguatera because a generally accepted test for toxic fish is not available. Warnings and advisories concerning the hazards of ciguatera and the risks of consuming particular species of fish from ciguatera areas are issued by states. Active control based on regulation of fishing dangerous species, supported by testing suspect fish at dockside or on board the catching vessel to detect and reject ciguatoxic fish, is proposed. Increased education of the consuming public, sports fishers, and health professionals on the hazards and symptoms of ciguatera is also recommended.

Scombroid poisoning reportedly caused about the same number of outbreaks as ciguatera but was much more widespread in occurrence. Tuna, mahimahi (dolphin), and bluefish were implicated as the major cause of scombroid poisoning in the United States. The disease is generally mild and self-resolving, and symptoms can be ameliorated by antihistamine drugs. Because the histamine that causes scombroid poisoning is produced after the fish have been caught as a consequence of improper temperature control, the disease can be prevented by rapidly cooling fish after capture to 10°C or lower and holding them at or below this temperature at all times before cooking and eating. A system based on the Hazard Analysis Critical Control Point would ensure this for commercially handled fish, but the education of subsistence and recreational fishers is also necessary.

Paralytic shellfish poisoning was reported as a minor cause of seafood-borne illness in 1978-1987 with only two deaths. This is a remarkable record in view of the annual occurrence of toxic situations among shellfish on both the East and the West coasts of the United States and indicates that current control measures applied by coastal states are highly effective. However, the increasing occurrence of toxic dinoflagellate blooms and changing eating practices among some sectors of the consuming public require increased surveillance and the development of more rapid and simple tests for toxic shellfish.

Although other natural seafood intoxications have not been reported recently in U.S. consumers (except for an outbreak of neurotoxic shellfish poisoning in North Carolina in 1987), the potential for their occurrence either from domestically produced seafoods or from imports is real. Increased vigilance concerning imported products, based on a requirement for certified nontoxicity, is recommended. Moreover, both state and federal laboratories should be prepared to test for these "other" toxins, and procedures should be in place to deal with outbreaks.

INTRODUCTION

The toxic diseases from fish and shellfish of importance to American consumers include ciguatera, scombroid fish poisoning, paralytic shellfish poisoning, neurotoxic (brevetoxic) shellfish poisoning, puffer fish poisoning, diarrhetic shellfish poisoning, and amnesic shellfish poisoning (Hughes and Merson, 1976; Mills and Passmore, 1988; Ragelis, 1984; Todd, 1989). In all cases, illness is due to ingestion of tissues containing heat-resistant toxins that are not destroyed by normal cooking and whose presence is undetectable by organoleptic means. Except for scombroid poisoning, toxins usually accumulate in fish or shellfish through the food chain, so that the fish or shellfish are toxic at the time of harvest. Scombroid poisoning is caused by bacterial-induced chemical changes resulting from mishandling of fish after capture, which is more readily susceptible to human control (Taylor, 1986).

Fish poisoning, principally ciguatera and scombroid fish poisoning, was responsible for 17.8% of all confirmed food-borne disease outbreaks listed by the Centers for Disease Control (CDC) in 1978-1987. Reports were approximately evenly split between the two principal toxicoses: 179 ciguatera outbreaks involving 791 cases, and 157 outbreaks of scombroid with 757 cases. However, as noted elsewhere in this report, CDC data are highly skewed, in this case due to the limited area within which ciguatera occurs, which enhances the visibility of this disease, and to the different symptoms associated with scombroid poisoning. Thirteen outbreaks of paralytic shellfish poisoning (PSP), the most dangerous of the intoxications, were reported and involved 134 cases, most of which (94) were from two large California outbreaks in 1980. No cases of puffer fish or diarrhetic shellfish poisoning were reported to CDC in this period. The actual incidences of cases of ciguatera and PSP with milder symptoms are probably higher than indicated due to underreporting, as evident from a comparison of CDC data with those obtained in incidence studies in defined geographical areas (Mills and Passmore, 1988; Morris et al., 1982b; Nishitani and Chew, 1988).

TABLE: Illness Due to Natural Seafood Toxins in the United States Reported to CDC

	Ciguatera		Scombroid		PSP	
	Outbreaks	Cases	Outbreaks	Cases	Outbreaks	Cases
1978	19	56	7	30	4	10
1979	21	97	14	134	1	3
1980	15	52	28	151	5	116
1981	30	219	9	93	-	-
1982	8	37	18	58	0	5
1983	13	43	13	271	-	-
1984	18	78	12	53	-	-
1985	26	104	14	56	2	3
1986	18	70	20	60	-	0
1987	11	35	22	95	-	0
Total	179	791	157	757	13	137

SPECIFIC INTOXICATIONS

Ciguatera

Ciguatera is a clinical syndrome caused by eating the flesh of toxic fish caught in tropical reef and island waters. The toxin is believed to originate in a microscopic dinoflagellate alga *Gambierdiscus toxicus* that grows on reefs (Bagnis et al., 1980). However, other benthic algae have also been implicated. Fish eating the algae become toxic, and the effect is magnified through the food chain so that large predatory fish become the most toxic. The occurrence of toxic fish tends to be localized, but localization is not consistent and toxic fish may occur sporadically anywhere in a reef or island location (Engleberg et al., 1983). More than 400 species have been implicated in ciguatera poisoning (Randall, 1980), but the fish most commonly implicated include amberjack, snapper, grouper, barracuda, goatfish, and reef fish belonging to the Carrangidae.

In the United States, ciguatera occurs principally in Hawaii, Puerto Rico, the Virgin Islands, Guam, and Florida (CDC,

1989). A particularly high incidence was reported from Guam (Haddock, 1989), and a few cases have been reported in other states caused by fish shipped from Florida. Cases are frequently associated with travel to endemic ciguatera areas such as Hawaii and the Virgin Islands, and there is concern that many cases are not recognized by mainland U.S. physicians.

The disease affects both gastrointestinal and neurological systems (Bagnis et al., 1979; Morris et al., 1982a). Gastrointestinal symptoms, including diarrhea, nausea, vomiting, and abdominal pain, appear 3-5 hours after ingestion of the fish and are of short duration. Neurological symptoms begin 12-18 hours after consumption and may be moderate to severe; they commonly last for 1-82 days but may persist for several months. In rare cases, symptoms may last for years, with exacerbation associated with fish consumption or possibly alcohol (Halstead, 1967). Symptoms typically include hot-cold inversion (hot coffee tastes cold, ice cream tastes hot); muscular aches; tingling and numbness of lips, tongue, and perioral region; metallic taste; dryness of mouth; anxiety; prostration; dizziness; chills; sweating; dilated eyes, blurred vision, and temporary blindness. Paralysis and death may occur in a few extreme cases. Symptoms may be extremely debilitating, resulting in extended periods of disability. Intravenous mannitol may relieve acute symptoms (Palafox et al., 1988), provided it is given within several hours of consumption, with amitriptyline (Bowman, 1984) or tocainide (Lange et al., 1988) suggested for more chronic manifestations. There is considerable individuality in patient response (Engleberg et al., 1983).

TABLE: Fish Reported as a Vehicle of Ciguatera to CDC 1978-1987 Ordered by Frequency of Reported Involvement[a]

Common Fish Name	Genus or Family Name
Amberjack	*Seriola* species
Snappers[b]	Lutjanidae
Groupers	Serranidae
Goatfish	Mullidae
Po'ou	*Cheilinus* species

Jacks	Carangidae
Barracuda	Sphyrenidae
Ulua	*Caranx* species
Wrasse	Labridae
Surgeon fish	Acantharidae
Moray eel	Muraeinidae
Papio	*Trachinotus* species
Roi	*Cephalopolis* species
Rabbit fish	*Siganus* species
Parrot fish	Scaridae
Miscellaneous reef fish	

[a] See Halstead (1967) for a more complete listing of fish species involved in Ciguatera outbreaks.[b] These categories include a number of different species commonly referred to by the vernacular name. Not all species in each category are necessarily toxic.SOURCE: CDC (1981a-c, 1983a,b, 1984, 1985, 1989) and Haddock (1989).

Several toxic compounds have been isolated from ciguatoxic fish and from *Gambierdiscus*. The principal toxin called "ciguatoxin" is a small lipid-soluble polyether with a molecular weight of 1,112 (Scheuer et al., 1967); this toxin has been purified and its structure determined (Murata et al., 1990). Ciguatoxin (CTX) has a molecular formula of $C_{60}H_{88}O_{19}$ and is a brevitoxin type polyether, approximately 100 times more potent than terodotoxin. Ciguatoxin opens voltage-dependent sodium channels in cell membranes (Bidard, 1984), and studies with in vitro tissue preparations suggest that the toxin causes a nerve conduction block after initial neural stimulation. In animal models, low doses of ciguatoxin cause mild hypotension and brachycardia. Higher doses give a biphasic response with an initial brachycardia/hypotension followed by tachycardia/ hypertension; very high doses produce a phrenic nerve block with respiratory arrest (Gillespie et al., 1986). Another lipid-soluble neurotoxin from ciguateric fish is called "scaritoxin." This toxin has been shown to depress oxidative metabolic processes in rat brain and has a depolarizing action on excitable membranes.

Generally, the pharmacological action is close to that of ciguatoxin, and they may be related compounds (Legrand and Bagnis, 1984). Maitotoxin is a water-soluble toxin that may interfere with or modify calcium movement or calcium conductance in tissues. Other lipid-soluble toxins have been reported, but their structures and pharmacologic roles are not understood (Ragelis, 1984).

The reported incidence of ciguatera as indicated by CDC data is on the order of 15-20 outbreaks per year, involving 50 to 100 cases. Cases reported to the CDC occur almost exclusively in Hawaii, Puerto Rico, the Virgin Islands, and Florida (Table 4-3). However, these numbers appear to reflect significant underreporting. In a randomized, stratified community survey conducted in the U.S. Virgin Islands, the calculated incidence rate was 73 cases/10,000 population/year (Morris et al., 1982b). In Puerto Rico, 45 cases were reported to the Puerto Rico Poison Control Center in 1982. In an associated telephone survey, 7% of persons contacted reported that at least one family member had at one time had ciguatera (Holt et al., 1984).

In Miami, 129 cases of ciguatera were reported to the Dade County Health Department between 1972 and 1976, for an annual incidence of 5 cases/100,000 population; the actual incidence was estimated to be 10-100 times this figure (i.e., 50-500 cases/100,000 population) (Lawrence et al., 1980). An incidence rate of 234.9 cases/100,000 population was reported for the Marshall Islands during 1982-1987 (Ruff, 1989). The average annual incidence in Hawaii for 1984-1988 was only 8.7 cases/100,000 population, but this varied greatly from island to island: in 1988 the rate per 100,000 population was 3.2 on Oahu, 12.5 on Kauai, 11.1 on Maui, 33.9 on Hawaii (the largest island), and 7.5 for the state (Gollop and Pon, 1991). These data emphasize the striking regional nature of this disease and its very real importance as a cause of morbidity in endemic areas. There is some evidence from the Pacific region that changes in the reef environment due to construction or other underwater activities can cause an increase in the occurrence of ciguatoxic fish (Anderson et al., 1983; Ruff, 1989).

For the vast majority of U.S. consumers, the disease can be contracted only through consumption of fish imported from endemic areas. For residents of south Florida, the Caribbean, and Hawaiian or other Pacific islands, absolute safety depends on individual abstinence from eating reef fish. The risk may be greatly reduced by avoidance of particular species of fish from known "hot-spot" areas (Randall, 1980). However, hot spots may persist for extended periods of time or may change (Cooper, 1964; Halstead, 1967). Voluntary action by commercial fish distributors in some areas has been quite effective. Thus amberjack (*Seriola dumerili*, Kahala) is not sold commercially in Hawaii because of the known high incidence of ciguatoxic fish of this species. Other suspect species coming to market in Hawaii may be tested by the "stick test." Also, the Hawaii Department of Health (HDH, 1988) publishes a warning pamphlet on *Fish Poisoning in Hawaii* and periodically issues advisories on dangerous species and the areas from which they have been taken.

TABLE: Outbreaks and Cases of Ciguatera in the United States Reported to CDC 1978-1987

State	Outbreaks	Cases	Percentage of Outbreaks	Percentage of Cases
California	1	2	0.6	0.3
Florida	9	35	5.0	4.4
Guam[a]	60	117	56.0	48.0
Hawaii	144	560	80.0	71.0
Louisiana	1	6	0.6	0.8
Puerto Rico	13	73	7.0	9.0
Vermont	1	3	0.6	0.4
Virgin Islands	9	110	5.0	14.0
Washington	1	2	0.6	0.3
Total	179	791		

[a] From Haddock (1989). These data are not included in calculation of total outbreaks and cases.

In areas where reef fish are part of the regular diet of inhabitants or visitors, and particularly where significant quantities of fish are caught and consumed by recreational or small-boat fishermen, it seems unlikely that ciguatera can be completely prevented. The impossibility of detecting toxic fish by organoleptic inspection and the sporadic occurrence of such fish limit control options. The availability of a simple reliable test would greatly improve the situation. At present, the only ciguatera screening program in existence is that employed by the Tokyo Central Wholesale Fish Market in Japan. Hygiene inspectors examine incoming shiploads of fish that have originated in tropical island regions. Suspect specimens are removed for testing. Muscle extracts are prepared and tested on cats and mice for evidence of ciguatoxicity (Halstead, 1970). This is a lengthy and expensive screening technique that is impractical when dealing with large numbers of samples. A radioimmunoassay (RIA) was developed by Hokama and co-workers in Hawaii (Hokama et al., 1977) and then modified to a simpler, enzyme immunoassay (Hokama, 1985). The method has been further simplified to a "stick" test that has been used to screen fish landed in Hawaii and holds promise as a practical basis for control (Hokama et al., 1989b).

However, even this test costs $1 to 2 per fish, and it would not be possible to test each reef fish landed. Kits are being developed for use by sports fishers that could partly resolve the cost problem. In any case it would be desirable to limit testing to high-risk fish. Research is needed into methods for predicting the development of ciguateric conditions in reef fishing areas, perhaps by assessing *Gambierdiscus* or other toxigenic microorganism populations and somehow closing such areas to fishing when the risk is high. Reef closure would probably be feasible in discrete Pacific islands where there is limited movement of fish from one reef area to another. However, this might not be the case in the Caribbean where fish movement between reefs is easier and more common. Obviously, there should be some follow-up on the reports of a relationship between reef disturbance and increased occurrence of ciguateric fish because this may result from human activity that can be controlled (Gollop and Pon, 1991).

Estimates of the economic consequences of ciguatera are not easily made. However, they are significant for island communities largely dependent on tourism. Ragelis (1984) quoted an estimate of an annual loss to fishermen in the

Caribbean region and Hawaii of $10 million as a result of restricted fishing, but this may be low.

In summary, the risks of contracting ciguatera fish poisoning are low for most consumers of seafood in the mainland United States. Risks are much higher in Hawaii, other Pacific islands, Puerto Rico, and the Virgin Islands, with more moderate risks in areas such as Miami that border endemic zones. For mainland consumers, protection could be afforded by strict control of imports and intrastate shipments. However, such an approach may be unnecessarily severe and does not address the much more significant problems that exist in endemic areas. A more reasonable (and potentially cost-effective) approach would be to emphasize development of an inexpensive but reliable assay for ciguatoxic fish, similar to the stick test proposed by Hokama (1990). The stick test measures ciguatoxin and polyether compounds including okadaic acid (Hokama et al., 1989a).

In a recent outbreak in Hawaii due to Philippine fish, the test was positive for a fish that was then shown to contain palytoxin (Kodama et al., 1989). It has been suggested that palytoxin, previously reported from parrot fish and crab in Japan, is one of the toxins "under the rubric of ciguatera" (Kodama et al., 1989). Obviously, this whole area needs further research, particularly because of the concern over false-positive results from the stick test.

If such an assay were widely available, it might be applied both by regulation and voluntarily to reduce the incidence of disease in endemic areas, particularly if consumers were knowledgeable and insisted on purchasing only fish that had been screened for toxicity. Similarly, interstate shipment and imports of potentially high-risk fish (grouper, jack) could be restricted to fish certified to be nontoxic. This is particularly important in view of the increased production and export of fish to the United States from Pacific islands and reef fishery areas of Southeast Asia, such as the Philippines (Miller, 1991).

Scombroid (Histamine) Fish Poisoning

Scombroid intoxication results from ingestion of fish containing high levels of free histamine. Initially, the disease was associated with consumption of scombroid fish such as tuna, mackerel, bonito, and saury. More recently, other types of fish have been identified as causing the intoxication, including mahimahi, bluefish, jack, mackerel, amberjack, herring, sardine, and anchovy. In the United States, scombroid fish poisoning has been caused dominantly by mahimahi, tuna, and bluefish (CDC, 1989).

Scombroid food poisoning has a wider geographic occurrence in the United States than ciguatera, with incidents reported from 45 states during 1978-1988.

TABLE: Fish Reported to CDC as Vehicles of Scombroid Poisoning in the United States, 1978-1986[a]

Common Name	Genus	Reported Outbreaks
Mahimahi	*Coryphaena*	55
Tuna	*Thunnus*	41
Bluefish	*Pomatomus*	13
Salmon (raw)	*Oncorhynchus*	2
Marlin	*Makaira*	1
Mackerel	*Scomber*	1
Blue Ulua	*Caranx*	1
Opelu	*Dicapterus*	1
Redfish	*Sebastes*	1

[a] Data not available for 1987.

The highest number of outbreaks (45) and cases (171) occurred in Hawaii, but mainland states reported a total of 111 outbreaks and 582 cases. This reflects the fact that the disease, although associated with warm ambient temperatures, is not due solely to tropical or subtropical species of fish. Thus, the risk of scombroid poisoning is widespread among fish-eating consumers.

Fortunately, the disease is mild, of short duration, and self-resolving without any sequelae in the vast majority of cases. Moreover, because the toxic condition is a consequence of improper handling or storage of the fish and there are effective testing methods to identify toxic fish, control and prevention are possible. The mildness and transient nature of scombroid poisoning make it likely that this disease is underreported.

Fish imported to the United States from warmwater countries, particularly mahimahi, have been implicated as a cause of scombroid poisoning; this reflects both the high ambient water and air temperatures in the originating area, and the poor handling conditions on boats and in markets permitting growth of the bacteria that convert histidine to histamine.

The disease is correctly described as histamine poisoning (Taylor, 1986); it includes gastrointestinal, neurological, hemodynamic, and cutaneous symptoms such as nausea, vomiting, diarrhea, cramping, headache, palpitations, flushing, tingling, burning, itching, hypotension, rash, urticaria, edema, and localized inflammation. The most frequent symptoms are tingling and burning sensations around the mouth ("peppery tasting"), gastrointestinal complaints, and a rash with itching. The illness is generally mild and self-resolving, with rapid onset of symptoms and duration of only a few hours. Normally, treatment is unnecessary but antihistamine drugs will provide relief.

TABLE: Scombroid Fish Poisoning in the United States Reported to CDC, 1978-1987

State	Outbreaks	Cases
Alaska	3	17
Arizona	3	7
California	18 (12%)	69 (9%)
Connecticut	8	47
District of Columbia	1	3
Florida	1	20
Hawaii	45 (29%)	170 (23%)
Idaho	1	4
Illinois	3	35
Indiana	1	4
Kentucky	1	7
Maine	3	54
Maryland	1	10

Michigan	3	25
Minnesota	1	24
Nebraska	1	10
New Jersey	4	42
New Mexico	1	2
New York	30 (19%)	122 (16%)
North Carolina	1	10
Pennsylvania	2	4
Texas	2	11
Vermont	3	6
Virginia	2	10
Virgin Islands	1	5
Washington	16	35
Wisconsin	1	1
Total	157	757

The histamine is produced in the fish flesh by decarboxylation of free histidine, which is naturally present at high levels in species of fish implicated in scombroid fish poisoning (Lukton and Olcott, 1958). The production of histamine is due to the action of histidine decarboxylase, an enzyme produced by bacteria growing on the fish. Histidine decarboxylase production is not widespread among bacteria and is found principally among species of the Enterobacteriaceae, *Clostridium, Lactobacillus* (Taylor, 1986), and possibly *Vibrio* (Van Spreekens, 1987).

The enteric bacteria *Morganella morganii, Klebsiella pneumoniae*, and *Hafnia alvei* have been isolated and identified from fish implicated in histamine poisoning (Havelka, 1967; Kawabata et al., 1956; Taylor et al., 1979). Other enteric bacteria, *Clostridium perfringens*, and halophilic vibrios have also been reported, but *M. morganii* and *K. pneumoniae* are most frequently implicated. These organisms are not commonly isolated from living fish and may be added during catching and handling (Taylor, 1986).

Bacteria must grow to a large enough population for significant production of histamine to occur. These are mesophilic bacteria that require temperatures higher than 15°C. In tropical areas of the world, fish temperatures at capture frequently exceed 20°C, and on small vessels it is not unusual for fish to be held on deck at even higher temperatures for several hours. Histamine production is optimal at 30°C (Arnold et al., 1980). Once a large population of bacteria has been established, residual enzyme activity continues slowly at refrigeration temperatures (0-5°C) though bacterial growth ceases.

Thus, histamine production in fish is a consequence of improper handling and storage of fish after capture. Indeed, histamine content may be used as an index of spoilage in certain fish. The Food and Drug Administration (FDA) considers a level of 20 milligrams (mg) of histamine per 100 grams (g) of flesh, or 200 parts per million (ppm), an indication of spoilage in tuna and 50 mg/100 g (500 ppm) an indication of hazard (Federal Register, 1982). This is close to the toxic dose estimate of 60 mg/100 g made by Simidu and Hibiku (1955). There is uncertainty regarding the threshold toxic dose because potentiators of toxicity are present in fish that lower the effective dosage compared with pure histamine.

The occurrence of scombroid fish poisoning in recent years, based on CDC reports, is between 12 and 20 outbreaks involving fewer than 100 cases per year (higher numbers were recorded in 1973, 1979, and 1980). This is, without question, a considerable underestimate because the illness is generally mild, passes rapidly with no after effects, and is thus not usually reported to health authorities.

Good chemical tests are available for histamine in fish flesh (Taylor, 1986), which has allowed FDA to set an action level for histamine in tuna at 50 mg/100 g of flesh. Above this level the fish is considered hazardous. Fish histamine poisoning is preventable by proper handling of fish at the time of capture and during subsequent storage, processing, and distribution. Fish should be chilled as rapidly as possible after capture by using ice, refrigerated seawater or brine, or mechanical refrigeration. Flesh temperature should be brought below 15°C and preferably below 10°C within 4 hours; this should be normal practice in commercial systems.

Histamine levels should be monitored routinely by the industry in susceptible species where proper prior handling cannot be ensured. The level at which testing is performed will depend on the species and product form (e.g., tuna for canning, hot smoked mackerel). In the United States, the highest-risk fish commercially is probably imported fresh or frozen fish from tropical areas. High histamine levels may be present in such fish when other overt signs of spoilage (bad odor, discoloration) are absent.

Imported fish should be subject to controls. Domestically caught species in normal commercial channels are probably less of a problem because of the widespread use of ice and refrigeration. Bluefish and sport caught tuna or mackerel present a more intractable control problem because they are caught by individuals and either do not enter commercial channels or do so in an unconventional way.

Some local or state control may be possible where licensed charter fishing boats are involved, perhaps by requiring that adequate facilities are provided for rapid chilling of fish and for their storage in a chilled state until landed. In the absence of a simple litmus test, control for most sports fishers and their families depends on education. States should be encouraged to provide advisory bulletins to sports fishers. However, it should be emphasized that this is a mild disease that is neither long lasting nor life threatening, and that symptoms can be relieved quickly by antihistamines.

Paralytic Shellfish Poisoning (PSP)

Paralytic shellfish poisoning results from ingesting bivalve molluscs (mussels, clams, oysters, scallops) that have consumed toxigenic dinoflagellates (Halstead and Schantz, 1984; Schantz, 1973). The toxins are assimilated and temporarily stored by the shellfish. In the United States, PSP is a problem primarily in the New England states on the East Coast and in Alaska, California, and Washington on the West Coast. Very few outbreaks have occurred in other areas of the United States from shellfish harvested in coastal states, reflecting the effectiveness of current testing and control measures for commercially produced shellfish. Most disease incidents involve mussels, clams, and scallops gathered and eaten by recreational collectors often from closed areas.

The CDC listed 12 outbreaks involving 134 people with one death during 1978-1986. The outbreaks occurred in Alaska, California, Massachusetts, Tennessee (due to mussels from California), and Washington.

The Northeast Technical Support Unit (NETSU) compilation shows a total of 282 cases for the period, including cases from Maine, Alaska, and Massachusetts. A report by Nishitani and Chew (1988), based on data from the West Coast, lists cases as follows: 68 from Alaska, 98 from California, 1 from Oregon, and 12 from Washington. Thus, there is some evidence of underreporting of cases to CDC. Although PSP is an extremely dangerous disease that can cause death, there is reason to believe that mild cases due to consumption of marginally toxic clams by recreational diggers are never reported to health authorities or are misdiagnosed.

Paralytic shellfish poisoning is potentially life threatening because the toxins involved are among the most poisonous known. Symptoms are neurological and normally appear within an hour of eating toxic shellfish; in nonlethal cases they usually subside within a few days.

Symptoms include tingling, numbness, and burning of the lips and fingertips; ataxia; giddiness; staggering; drowsiness; dry throat and skin; incoherence; aphasia; rash; and fever. In severe cases, respiratory paralysis occurs, which can cause death usually during the first 24 hours, so that the prognosis for recovery is good for patients surviving this period. No antidote is known, but respiratory support is given when paralysis occurs.

There are no sequelae, and patients recover completely. Immunity is not conferred by a poisonous episode and multiple incidents can occur.

TABLE: PSP Incidents in the United States Reported by CDC, 1978-1985

	Outbreaks	Cases	Deaths
Alaska	2	7	—
California	2	94	2
Massachusetts	1	15	—
Tennessee	1	5	—
Washington	7	16	—

TABLE: PSP Incidents in the United States Listed Elsewhere

	NETSU Cases	Cases Reported by Nishitani and Chew (1988)[a]
Alaska	54	75
California	132[b]	98
Maine	73	—
Massachusetts	41	—
Oregon	-	1
Tennessee	5	—
Washington	15	12

[a] 1978-1987: West Coast states only.

[b] 1969-1980.

The cause of PSP is a complex of toxins known as saxitoxins because all can be considered forms or derivatives of saxitoxin, whose structure was reported by Schantz et al. (1975). The 12 most commonly encountered include saxitoxin, neosaxitoxin, gonyautoxins (I, II, III, IV), B1, B2, C1, C2, C3, and C4, which vary in their toxic effects on mice (Boyer et al., 1978; Shimizu and Hsu, 1981). Saxitoxin, neosaxitoxin, and gonyautoxins II and III are roughly equal in toxicity, whereas the others are somewhat weaker (Hall and Reichardt, 1984). In the United States, the toxigenic dinoflagellates of importance are *Gonyaulax catenella* and *G. tamarenses*.[1] the first being most dominant on the West Coast and the second on the East Coast (Taylor, 1988). Strains of these microorganisms develop characteristic toxin profiles that usually contain six to eight saxitoxins. Shellfish feeding on blooms of these *Gonyaulax* ingest all toxins but seem to selectively retain or biologically modify some derivatives because the toxin profiles in clams or scallops may differ from those of the *Gonyaulax* on which they have been feeding (Schantz et al., 1975; Sullivan et al., 1983).

The saxitoxins are neurotoxins that act by blocking the flow of sodium (Na^+) ions through the sodium channels of nerves, thereby interfering with signal transmission. The lethal dose for humans is 1-4 mg expressed as saxitoxin equivalents (Schantz, 1986), and the FDA action limit is 80 micrograms (μg) of toxin per 100 g of shellfish tissue.

The classical method for analysis of saxitoxins is by mouse bioassay (Schantz et al., 1958), a standardized procedure in which the strain, size, and condition of the mouse are all important. Suitable groups of mice are challenged by injection of toxin extracts, and their responses are compared with those of mice injected with known toxin at different concentrations. This method is used by state laboratories in surveillance activities. The chemical analytical methods that have been used, mostly in research, include column separation, thin-layer chromatography, and fluorescent assays either directly or after separation by high-performance liquid chromatography (HPLC). The HPLC method is quite rapid and has been considered as a possible replacement of the mouse bioassay (Sullivan and Iwaoka, 1983). Recently, immunoassay methods have been developed and reported. These involve rabbit serum antibody preparations and monoclonal antibodies, and both radioimmunoassays and enzyme-linked immunoabsorbent assays (ELISA) have been used (Chu and Fan, 1985). However, so far these methods are not acceptable for regulatory use.

Blooms of the toxigenic dinoflagellates *Gonyaulax catenella* (West Coast) and *G. tamarensis* (East Coast) occur several times each year, primarily from April to October along the U.S. West Coast, Alaska, and the East Coast from Long Island Sound through Maine. Similar blooms occur off British Columbia and the Canadian Maritime Provinces. The occurrence of blooms is not predictable with the present state of knowledge. When these blooms occur, shellfish become toxic and remain toxic for several weeks after the bloom subsides. Some species of bivalve molluscs in a few areas remain constantly toxic (e.g., butter clams in parts of Washington State and in Alaska). Protection of the consumer is primarily the responsibility of the state in which potentially toxic shellfish originate and is achieved by closure of shellfish harvesting in affected areas.

Closure may be absolute for certain species over a long period (e.g., ocean coast mussels in California) or temporary (e.g., most hard-shell clam operations in Washington and Oregon) (Nishitani and Chew, 1988). Because clam digging and oyster gathering from public beaches are popular recreational activities in coastal states, the authorities issue warnings through the media when dangerous conditions exist and post warning notices (multilingual in the West) on public beaches. Each of the affected states has a surveillance system involving regular sampling and testing of shellfish from different areas during the season for blooms (April to October). Toxicity tests are normally run in state public health laboratories by using the standardized mouse bioassay (AOAC, 1984), whereas the collection of samples and posting of beaches are frequently the responsibility of local (e.g., county) authorities. Commercial producers are required to submit samples for testing, and many voluntarily send shellfish to the state laboratories.

This system, although expensive for the states, has served the U.S. consumer well, which can be seen from the very small number of outbreaks reported to CDC in recent years. In 1977-1986, these involved only twelve outbreaks with 134 cases. Only two outbreaks, both due to mussels from the same source (California), were of commercial origin. All other outbreaks were among recreational clam diggers (shellfish gatherers). Despite the seriousness of the intoxication, there have been only two deaths in the last 10 years. West Coast state records show a somewhat higher number of outbreaks, which have typically involved recreational shellfish gatherers but with no serious consequences (Nishitani and Chew, 1988).

The present system of testing and control by the states appears to provide adequate protection to the consumer of domestically produced shellfish.

Neurotoxic Shellfish Poisoning (NSP)

Neurotoxic shellfish poisoning (sometimes referred to as brevetoxic shellfish poisoning, BSP) is caused by ingesting shellfish that have fed on the red tide organism *Gymnodinium breve* (formerly *Ptychodiscus brevis*). Red tides occur sporadically in the Gulf of Mexico and off the coast of Florida (Baden et al., 1984). They may be carried north in the Gulf Stream, occasionally affecting the coastline of adjacent states. The dinoflagellate blooms are easily observed as a red coloration of seawater, and the organisms can be detected microscopically. Red tides usually cause massive fish kills, and the carcasses wash ashore. Irritant aerosols are produced by wind and wave action, which may cause respiratory distress.

Filter feeding molluscs ingest the dinoflagellates and retain the toxin in their tissues for some time. No cases of NSP were reported to CDC in the period 1978-1986, but five cases were reported from Florida in 1973-1974 according to NETSU. During a red tide incident in North Carolina in 1987-1988, 48 persons became ill with NSP (Tester and Fowler, 1990). Reports of respiratory irritation affecting people in coastal areas were received on October 29 and 30, 1987; oyster harvesting was closed on November 2. However, 35 of the reported cases occurred prior to the ban on shellfish harvesting which lasted from 3 1/2-6 months, depending on location. The basis for closure was the occurrence of more than 5,000 *G. breve* cells per liter of seawater, and reopening of harvest was dependent on demonstrated absence of brevetoxin in 100-g samples of shellfish meat based on a mouse bioassay (DNR, 1985; FDA, 1989). Notwithstanding this incident, which occurred as a result of the first bloom of *G. breve* reported in North Carolina, the general record in recent years suggests that the surveillance and closure systems operated by the states are indeed effective.

Symptoms resulting from the ingestion of shellfish containing brevetoxins include tingling and numbness of the lips, tongue, throat, and perioral area; muscular aches;

gastrointestinal upset; and dizziness. The intoxication is usually not fatal. Onset is rapid and symptoms subside within a few hours or days at most. There is no antidote.

The symptoms appear to be due to two brevetoxins produced by *G. breve* that bind to nerve cells (Baden et al., 1984). They may be analyzed chemically, but this is not done routinely. Identification of a dangerous condition is readily made by observation of red tide conditions, including characteristic fish kills, and of the organisms themselves in the water. Local authorities then routinely close shellfish harvesting to industries and the public.

Diarrhetic Shellfish Poisoning (DSP)

Diarrhetic shellfish poisoning is caused by ingestion of mussels, scallops, or clams that have been feeding on *Dinophysis fortii* or *D. acuminata* and other species of *Dinophysis* and possibly *Prorocentrum* (Edler and Hageltorn, 1990; Yasumoto and Murata, 1990). There have been no confirmed outbreaks in the United States, but the disease is common in Japan and has become a problem in Europe. One confirmed DSP episode occurred in Canada in 1990.

Symptoms include diarrhea, nausea, vomiting, and abdominal pain. Onset occurs from 30 minutes to a few hours after eating toxic shellfish, and the duration is usually short with a maximum of a few days in severe cases. The disease is not life threatening (Yasumoto et al., 1984).

At least five toxins have been isolated from dinoflagellates and shellfish. Okadaic acid is most commonly encountered in Europe where *D. acuminata* is the usual agent, and mixtures of okadaic acid, dinophysistoxins, and pectenotoxins are detected in Japanese cases usually involving *D. fortii* (Yasumoto and Murata, 1990). There is a mouse bioassay for the toxins.

For the U.S. consumer, DSP would appear, at present, to be a hazard only for imported products and should be controllable by import regulation. Shellfish should be imported only from countries with whom the United States has a memorandum of understanding (MOU). Testing for shellfish toxins should be part of the general practice under the MOU. Nevertheless, because *Dinophysis* does occur in U.S. coastal waters, regulatory agencies in the United States should be alert for the possibility of an outbreak (Freudenthal and Jijina, 1988).

Puffer Fish Poisoning (PFP)

Puffer fish poisoning results from ingestion of the flesh of certain species of fish belonging to the Tetraodontidae (Halstead, 1967). The toxin involved is called tetrodotoxin and was originally believed to be a true ichthyosarcotoxin produced by the fish itself. The toxicity of poisonous puffers fluctuates greatly (Halstead, 1988). Recent observations that cultured puffer fish are atoxic has supported a food chain origin for the toxin, but this has not yet been confirmed (Mosher and Fuhrman, 1984). It has recently been shown, however, that certain common marine vibrios can produce a form of the toxin (Narita et al., 1987), and because vibrios occur as part of the microflora of puffer fish, they may be implicated in toxicity development (Sugita et al., 1989).

Puffer fish poisoning has not been reported in mainland United States in recent years, but incidents were reported in the past. Seven cases were reported in Florida between 1951 and 1974, including three fatalities (Benson, 1956; Hemmert, 1974). They appear to have been caused by the consumption of locally caught species of *Sphoeroides*.

The common puffer fish, *Arothron hispidus*, has been implicated in at least seven fatalities in Hawaii (HDH, 1988). There are 20-100 deaths from fugu poisoning in Japan each year, where various species of puffer fish are eaten as a delicacy; this occurs despite very stringent controls imposed by Japanese authorities on the marketing and restaurant preparation of the dish (Ogura, 1971).

The symptoms of puffer fish poisoning are similar to those described for paralytic shellfish poisoning, including initial tingling and numbness of lips, tongue, and fingers leading to paralysis of the extremities; ataxia; difficulty in speaking; and finally, death by asphyxiation due to respiratory paralysis. Nausea and vomiting are common early symptoms. The similarity in symptoms is not surprising because tetrodotoxin.

although chemically different from the saxitoxins, also blocks sodium channels. No antidote has been identified for tetrodotoxin and treatment is supportive. The toxicity of tetrodotoxin is similar to that of saxitoxin, and 1-4 mg constitutes a lethal dose for humans.

There is disagreement concerning the toxicity of U.S. Atlantic puffer fish. A recent advisory from the National Oceanic and Atmospheric Administration (NOAA, 1988) describes the northern puffer (*Sphoeroides maculatus*) as nontoxic and notes that the fish were marketed along the Atlantic Coast as "sea squab" during World War II. However, Hemmert (1974) shows a table indicating that the viscera, skin, and some flesh of *S. maculatus* caught in the Atlantic were toxic (Lalone et al., 1963). Larson et al. (1959, 1960) also reported that *S. maculatus* is toxic. From the West Coast, Goe and Halstead (1953) reported that the Pacific species *S. annulatus* is often toxic. The species *Arothron hispidus* has been implicated in at least seven fatalities in Hawaii. The wholesaling, preparation, and selling of puffers as food in Japan, even under the most rigid public health conditions by trained and certified puffer cooks, has not eliminated the danger of eating these fish. The fugu (puffer) still remains a major cause of fatal food intoxications in Japan. In brief, eating poisonous puffers is at best a game of Russian roulette. All of the U.S. puffers may be potentially toxic. There are too many variables in the puffer business, and sale of these should be prohibited in the United States. This subject has been documented and discussed at great length by Halstead (1967, 1988).

In view of these reports, it would seem prudent to exclude puffer fish, whether domestic or imported, from U.S. commercial channels at least until a proper assessment is made of the extent of risk they may present. The FDA has recently approved the importation of the Japanese puffer for fugu restaurants in the United States. Even though very strict requirements have been imposed in an attempt to ensure that the fish are nontoxic, the continuing Japanese experience should raise questions concerning the safety of this process for the U.S. public (Halstead, 1988).

Amnesic Shellfish Poisoning (ASP)

Amnesic shellfish poisoning has been proposed by Todd (1989) as a name for the syndrome caused by domoic acid. This severe disease has been identified only in a series of outbreaks in Canada in November and December 1988 involving 103 people.

The toxin is present in some varieties of the diatom *Nitzschia pungens* and accumulated in mussels and clams in Atlantic Canada during a period of blooms of the diatom. Symptoms included vomiting, abdominal cramps, diarrhea, disorientation, and memory loss (Perl et al., 1988; Teitelbaum et al., 1990). Short-term memory loss was the most persistent symptom and lasted over a year in several cases. Autopsies on three fatalities showed necrosis of the hippocampus. The disease is particularly severe among older people, some of whom died in the Canadian outbreak.

Canadian authorities now analyze mussels and clams for domoic acid and enforce closure of beds when levels in excess of 20 μg/g are detected in their tissues (Gilgan et al., 1989).

Clearly, this is a toxin to be considered in U.S. testing regimes, and there should be close cooperation between U.S. and Canadian regulatory agencies on the movement of imported Canadian shellfish into the United States. *Nitzschia pungens* and *N. pseudodelicatissima* reportedly occur in northern U.S. and Canadian waters, and there is potential for development of toxicity in shellfish growing in these areas. States in the northeastern United States are now testing mussels for domoic acid.

Other Toxins

There are sporadic reports of other intoxications from seafoods from time to time (Halstead, 1988; Wekell and Liston, 1982), but these have not been investigated sufficiently to identify the toxic agent. As noted earlier, the somewhat variable symptoms defined as ciguatera and the reported association of polyether substances and palytoxin (Hokama et al., 1989a; Kodama et al., 1989) in fish implicated in such cases raise questions about the toxicity of reef-associated fish.

One well-defined syndrome reported to occur in Hawaii is "hallucinogenic fish poisoning." This illness follows consumption of mullet and a number of reef fish and occurs seasonally, usually during summer months. Hallucinations, insomnia, intense dreaming, weakness, and burning of the throat are common soon after eating the fish (Halstead and Schantz, 1984). Terrifying nightmares have been reported and constrictive chest pains occur.

The condition is short-lived and self-resolving (Halstead, 1988; HDH, 1988). There does not seem to be any analytical test for this toxin.

CONCLUSIONS AND RECOMMENDATIONS

Three of these are of direct significance to the U.S. consumer: ciguatera, scombroid poisoning, and paralytic shellfish poisoning. Of these three, PSP, which has potentially the most severe health consequences, is well controlled by state surveillance and harvest closure practices.

Ciguatera, for which the largest number of cases is reported, has a major public health impact in Hawaii, Guam, and Caribbean island communities and a small effect in Florida. Prevention of ciguatera can be ensured only by interdiction of the supply of potentially toxic tropical reef fish to the U.S. consumer.

This is theoretically possible through banning imports to the U.S. mainland of fish known to become ciguatoxic and through strict control of fishing in dangerous areas by fishery management agencies, accompanied by rejection of suspect species at the point of landing. Such action would probably be unacceptable in the island states and possessions, where local fishing provides essential employment and is closely tied to the tourist industry.

Furthermore, blanket rejection of such species as groupers, which are mostly nontoxic, would greatly reduce consumer choice and adversely affect the income of fishermen and others in areas remote from the toxin problem. Fortunately, current research at the University of Hawaii provides good promise of early development of a simple reliable test for ciguatoxic fish.

Such a test is urgently needed to enable selective rejection of toxic fish by testing either on board the fishing vessel or at dockside. On a longer-term basis, research should be directed toward the prediction of developing toxic conditions so that closure of fishing areas can be applied before human intoxications occur.

To businesses supporting such activity (charter boats, gear suppliers, etc.) appear to be the only available solution. However, because imported mahimahi was reportedly responsible for 47% of identified scombroid poisoning, embargoing this fish could have a significant effect.

The other intoxications discussed in this chapter are rare and apparently under control in the United States (e.g., NSP and PFP), or not reported as a cause of sickness here (e.g., DSP and ASP).

Nevertheless, agencies responsible for ensuring the safety of the U.S. food supply should maintain constant vigilance to avoid the importation of such problems. Appropriate tests should be sought and laboratories prepared for their use. Importers should be required to ensure that seafood products from countries where such intoxications have occurred are not toxic. This is best done by controlling imports through an MOU that would include provisions for toxicity testing.

Organoleptic inspection systems have little value in protecting the consumer from seafood intoxications. Toxic fish and shellfish usually look and smell perfectly normal. Protection of the consumer requires a multifaceted approach involving industry practices and regulations, control of harvest and distribution, and as a last resort, testing, seizure, and detention.

This requires action by states and local authorities from different departments (e.g., fisheries and health), and a national program involving a single federal agency would probably not be effective without extensive state involvement.

In the final analysis the most effective measure is likely to be education of the fish-eating public about which fish and shellfish may be naturally toxic. Except for scombroid poisoning, toxicity is a function of the normal feeding habits of wild animals and cannot be controlled. Thus, potentially toxic fish may enter the food supply.

Fortunately, the serious life-threatening intoxications are controllable so that most incidents of fish poisoning are of short duration and are self-resolving. Nevertheless, research aimed at the detection and elimination of toxic fish from the food supply and at methods of treatment for intoxications such as ciguatera that can have long-lasting and even disabling effects should be encouraged.

There is a need for educational materials to be made available to the fishing industry, public health workers, divers, and sports fishers. A number of popular handbooks have been published dealing with the potential health hazards caused by marine organisms (Halstead, 1959, 1990; Halstead et al., 1990).

However, much of this information does not reach regulatory, clinical public health, and poison control centers (Freudenthal, 1990). In dealing with this subject matter, it is essential that the educational materials be fully illustrated, preferably in color. More charts and informational pamphlets are required.

In most cases toxic shellfish are not detectable by organoleptic means. It is therefore important that practical chemical or biological tests, specific for the detection of the toxins, be developed. Although not all the listed diseases are problems in the United States, seafood inspectors and processors should always be aware that toxigenic dinoflagellates, or other microorganisms producing toxins that get into fish and shellfish, may become established in areas where fish and shellfish are harvested for U.S. consumption. An example: diarrhetic shellfish poisoning is not a problem in the United States, but the dinoflagellate that produces the toxin may become established in shellfish areas that supply U.S. markets.

The committee recommends the following:

Fish of species reported by health authorities to have caused ciguatera, which are to be imported to the United States from regions of high ciguatera incidence, should carry certification of nontoxicity.

States and territories in which ciguatera is a problem should license marine sports fishers and, at the point of issuance of the license, issue clear and specific warnings regarding the dangers of ciguateric fish. Pamphlets on poisonous fish should be generally available to the public in areas where ciguatera is endemic.

When feasible, reef fishing should be closed in areas where ciguatoxic fish are present. This closure should apply to sports fishers as well as to commercial vessels.

Research should be accelerated on the development of simple, rapid tests for toxicity such as the Hokama stick test. Research should also be directed toward analysis of the events leading to the appearance of toxic fish in particular reef environments, with the objective of developing predictive indices that can be used to close areas to fishing before human intoxications occur.

All imported fish of species known to be a cause of scombroid poisoning should be certified as having histamine levels of less than 20 mg/100 g of fish. This should be controlled by routine lot testing.

Vessels fishing potentially scombrotoxic species should be required to maintain time/temperature records to ensure proper cooling and refrigerated storage of fish.

Similar temperature records should be maintained for such species during processing and shipment on land.

Advisory leaflets describing the causes of scombroid poisoning and providing advice on how to handle fish to minimize risk of the disease should be made widely available to sports fishers who target potentially scombrotoxic species.

Research to develop a rapid field test for PSP toxicity in shellfish should be strongly encouraged and supported. Such a test could be applied directly by commercial growers and recreational shellfish gatherers. Nevertheless, state agencies should continue to monitor the PSP condition of local shellfish.

The consumption of puffer fish should be strongly discouraged, and their importation to the United States should be banned.

Regulatory agencies should maintain awareness of potential toxin problems, such as diarrhetic shellfish poisoning and amnesic shellfish poisoning, and their technical personnel should be trained and equipped to run definitive analyses on these and other toxins. Shellfish should be imported only under an MOU that includes a provision for toxicity testing.

In view of the complexity of seafood intoxications, the federal government should establish or support two to three centers of research into such toxins to enlarge understanding of the phenomena, provide possible remedies, and develop particular tests.

Because of the highly localized impact, primary responsibility for control of seafood toxins should reside at the state level, with funding, quality control, and specialist assistance from a federal seafood safety agency.

6

Production, Distribution, and Consumption

Consumption of seafood has increased over the last decade, without a concomitant increase in reported illness. This increased consumption trend is expected to continue both for prepared and for fresh or frozen varieties. The 1989 consumption figure was 15.9 pounds of edible meat per person per year. Total commercial landings were a record 8.5 billion pounds in 1989, and imported edible products totaled 3.2 billion pounds. The majority of the seafood supply was harvested from wild populations. The aquaculture portion of this supply will probably increase.

A substantial amount of seafood (600 million pounds of finfish and 300 million pounds of shellfish) is caught recreationally. About 70% of commercially produced seafood in the United States is sold fresh or frozen. Canned seafood constitutes approximately 25%, and smoked/cured products 5%, of the seafood consumed. The United States exported 1.4 billion pounds of edible domestic fishery products in 1989.

The largest importer was Japan; Canada, the United Kingdom, France, and South Korea also provided good markets. The seafood harvest by industry is fragmented, diversified, seasonal, complex, and difficult to manage. Studies are needed to monitor changing consumption trends and patterns. The processing, distribution and merchandising of finfish and shellfish will require more emphasis to reduce cross-contamination. Attention must be given to aquaculture in order to produce high-quality, consistently available species. Attention must also be focused on the harvesting, handling, distribution, and preparation of recreationally harvested fish to ensure consumer safety. More emphasis should be placed on educating the industry and the consumer about safe handling practices that can reduce potential food-handling problems.

INTRODUCTION

As Americans become increasingly aware of the relation between diet and good health, the consumption of fishery products will most likely increase.

The consumer recognizes that fish and shellfish are nutritious and wholesome foods. They are perceived as an excellent source of high-quality protein, containing lipids with high levels of unsaturated fatty acids, and perhaps contributing to the enhancement of human health by reducing the risk of cardiovascular disease. Likewise, seafood is characteristically tender, easily digested, and a good source of many important minerals and vitamins (NRC, 1989).

Although the attributes of seafood attract a more health conscious consumer, they also enforce expectations for enhanced safety. Consumer consciousness of seafood safety issues has become, as a result, increasingly important. Pollution and environmental issues have further focused people's attention on contamination problems. Concurrent media coverage and public interest groups have heightened the demand for rigorous safety standards in the food industry (Haas et al., 1986; Newton, 1989). Unlike meat and poultry, which are derived from domesticated sources, the majority of the edible seafood supply in the United States, approximately 12.0 billion pounds including domestic landings and imports in round weight equivalents, was harvested from wild populations in 1989 (NMFS, 1990). The aquacultured portion of this supply is predicted to increase from both foreign and domestic sources (Redmayne, 1989), and recorded commercial landings are further supplemented by a growing portion of recreationally caught seafood destined for consumption.

Because the supplies of many seafoods are relatively small and regional, large numbers of individuals, using a variety of vessels that range from small boats to large factory ships, are involved. The seafood harvesting industry is highly fragmented. The diversity of the industry, the seasonal nature of fishing, the complexity of fish processing operations, and the substantial amount of seafood caught recreationally (600 million pounds of finfish and 300 million pounds of shellfish) make it difficult to manage and regulate these living resources (NOAA, 1990).

Both finfish and shellfish are subjected to contamination and cross-contamination in their natural habitat, as well as at any point during handling, processing, distribution, or preparation (Haas et al., 1986; Newton, 1989; NOAA, 1990). Seafood-borne illness has been reported due to natural toxins, microbial contamination, parasites, poor seafood handling, and chemical contaminants (CDC, 1981a-c, 1983a,b, 1984, 1985, 1989; FDA, 1989).

Because of the primary reliance on limited data-reporting systems via state departments of public health, and eventually the Centers for Disease Control (CDC), the extent of the public health risk due to cumulative exposure to microorganisms, natural toxins, and chemical contaminants cannot be assessed easily, especially in the context of total dietary exposure. Given this qualification, current data indicate a decrease in the reported incidents of illness from seafood relative to consumption.

The committee has critically examined and evaluated the degree of severity of illnesses, their significance, and the extent of possible health risks involved. Its findings are documented in subsequent chapters of this report.

DEMOGRAPHICS OF THE SEAFOOD INDUSTRY

In 1989, commercial and recreational fishermen harvested more than 8.5 billion pounds of fish and shellfish from U.S. waters, which includes edible and industrial products. More than 300 major species of seafood were marketed, reflecting the diversity of the resource base (NMFS, 1990). Over 4,000 processing and distribution plants handled the commercial products of the nation's 256,000 fishermen. Almost 95,000 boats and vessels constituted the fleet (NMFS, 1990).

Although commercial establishments are easily documented, the number of recreational fishermen and their support base are more difficult to quantify. Increasing numbers of anglers for fish from the nation's freshwater, estuarine, and marine waters are producing a growing share of the fresh and frozen seafood in today's diet. The number of recreational harvesters has been estimated to be in excess of 17 million individuals (NOAA, 1990).

Fresh and frozen seafood constitute about 70% of the product consumed in the United States. Canned seafood, particularly tuna, constitutes almost 25% of domestic consumption, and cured/smoked products account for the remaining 5% of per capita consumption.

FISHERY RESOURCES

Commercial landings (edible and industrial) by U.S. fishermen at ports in all the fishing states were a record 8.5 billion pounds (3.8 million metric tons) valued at $3.2 billion in 1989 (NMFS, 1990). This was an increase of 1.3 billion pounds (576,300 metric tons) in quantity, but a decrease of $281.8 million in value, compared with 1988. The total import value of edible fishery products was $5.5 billion in 1989, based on a record quantity of 3.2 billion pounds. Imports of nonedible (industrial) products set a record in 1989, with products valued at $4.1 billion, an increase of $676.1 million compared with 1988 (NMFS, 1990).

The trade deficit in fishery products has not declined. The dollar value of imports was higher in 1989 than in the previous year (NMFS, 1990). Canada is still the largest exporter to the United States, sending in more than 700 million pounds of fishery products in 1988. Ecuador was ranked second and Mexico third.

Whereas Canada ships finfish products, shrimp is the primary commodity exported by Ecuador and Mexico. Imports from Thailand and China are both increasing due to rising shrimp production from their expanding aquaculture systems.

On a worldwide basis, aquaculture is becoming a major new factor in seafood production. The cultivation of high-value species, popular in the U.S. market, is a major factor in import sourcing. China, for example, along with other Asian nations, is replacing South and Central American countries as a major shrimp supplier to the United States. Aquaculture is expected to determine much of the future fisheries growth, because wild stocks are nearing full utilization (NMFS, 1990; NOAA, 1990).

The total export value of edible and nonedible fishery products of domestic origin was a record $4.7 billion in 1989, an increase of $2.4 billion compared with 1988. The United States exported 1.4 billion pounds of edible products valued at $2.3 billion, compared with 1.1 billion pounds at $2.2 billion exported in 1988. Exports of nonedible products were valued at $2.4 billion. Japan continues to be America's best export customer. Over 700 million pounds of seafood was sold to the Japanese market, with salmon, crabs, and herring the primary commodities. Canada, the United Kingdom, France, and South Korea were also good markets in 1989, but the value of their imports was small, compared to Japan's purchase of West Coast products (NMFS, 1990).

Consumers in the United States spent an estimated $28.3 billion for fishery goods in 1989, a 5% increase from 1988 (NMFS, 1990). The total included $19.1 billion in expenditures in food service establishments (e.g., restaurants, carryouts, caterers); $9.0 billion in retail stores (for home consumption); and $181.7 million for industrial fish products. In producing and marketing a variety of fishery products for domestic and foreign markets, the commercial fishing industry contributed $17.2 billion in value-added dollars to the gross national product (GNP), an increase of 5% compared to 1988.

Consumption of fish and shellfish in the United States totaled 15.9 pounds of edible meat per person in 1989 (NMFS, 1990). This total was up 0.7 pound from the 15.2 pounds consumed per capita in 1988. Per capita consumption of fresh and frozen products registered a total of 10.5 pounds, an increase of 0.3 pound from the 1988 level. Fresh and frozen finfish consumption was 7.1 pounds per capita in 1989. Fresh and frozen shellfish consumption amounted to 3.4 pounds per capita, with canned fishery products at 5.1 pounds per capita, up 0.4

pound over 1988. The per capita use of all fishery products (edible and nonedible) was 62.2 pounds (round weight), up 2.8 pounds compared with 1988 (NMFS, 1990).

Although most of the fish and shellfish consumed is from commercial production, a significant share is caught recreationally. In 1990, the National Marine Fisheries Service (NMFS) estimated that 17 million marine anglers harvested more than 600 million pounds of finfish (NOAA, 1990). Although statistics are lacking, NMFS suggests that 200-300 million live pounds of molluscs and crustaceans was harvested by recreationalists. This catch represents 3-4 live pounds or about 1-1.5 edible pounds of domestic per capita consumption (Krebs-Smith, 1989), outside the commercial figure of over 15 edible pounds per person. The source, handling, and distribution of the recreational catch are just beginning to draw attention. Indeed, because recreational anglers are not regulated as food producers/manufacturers, there is concern about the use and distribution of this "recreational" resource.

Although it is difficult to give definite numbers for either the commercial or the recreational harvesting sector, some general observations can be made. Commercially, the trend is toward more efficient activity. Consequently, the number of participants in the commercial sector is decreasing. The commercial processing industry appears headed toward consolidation, with increased dependence on imported products and aquaculture. Recreational participation remains strong. Consumption data, as suggested by both the Department of Agriculture and the Department of Commerce, indicate a continued, if not expanding, harvest of sport caught fish and shellfish. More than 20% of all fresh and frozen seafood consumed in the United States may now be attributed to noncommercial harvest and distribution.

AQUACULTURE

Aquaculture is a rapidly growing mode of production in the seafood industry. Annual production of farmed fish and shellfish in the United States has grown 305% since 1980 (TFTC, 1988). The greatest production is of catfish (Sperber, 1989). Catfish production increased 31% from 1986 to 1987. According to the Catfish Institute, farm-raised catfish increased from 5.7 million pounds in 1970 to 295 million pounds in 1988 and were expected to exceed 310 million pounds in 1989 (Sperber, 1989). Salmon production in the Pacific Northwest and Maine totaled 85 million pounds in 1987. In addition, other fish that are farmed include trout, redfish, sturgeon, hybrid striped bass, carp, and tilapia, as well as shellfish and crustaceans such as oysters and crawfish. Crawfish production acreage has increased 145% to about 160,000 acres. Overall U.S. aquaculture production of fish and shellfish increased from 203 million pounds in 1980 to some 750 million pounds in 1987. It is estimated that by the year 2000, that figure will reach 1.26 billion pounds.

Large amounts of cultured fish and shellfish are also imported annually. Approximately one-half of the 500 million pounds of shrimp imported is cultured (Schnick, 1990); 143 million pounds comes from China and Ecuador, neither of which regulates the use of chemotherapeutic agents in culture. More than 40 million pounds of salmon is also imported annually, often from countries similarly lacking tolerance levels for residues. Of special interest are the use of chloramphenicol in shrimp culture and ampicillin in yellowtail culture (Hawke et al., 1987; Manci, 1990). The Food and Drug Administration (FDA) has not examined imported seafood for drug residues, and there is no information regarding levels that might be ingested (Schnick, 1990).

Aquaculture also produces fish used to stock recreational fishing areas. This procedure is under the control of government agencies that follow FDA regulations, use only approved drugs, and abide by legal withdrawal times.

CONSUMPTION TRENDS

Today's consumer is changing rapidly. Instead of single-income households, it is increasingly more common to have both man and woman working. The size of the family is decreasing. As many as one-fourth of all households are occupied by one person. This means more shoppers and diners, most with little time for home preparation (Davis, 1989).

Most adult men and women now work outside the home. In recent surveys, 7 out of 10 new home buyers noted that they will need two incomes to pay their respective mortgages. Nevertheless, the growth in two-income couples has generally created an increase in disposable income, but with little time to spend it. With as many as 50% of new mothers working outside the home within the first year of childbirth, it is easy to see the revolutionary changes taking place among families. The working mother or single dweller does not have the time to prepare meals in the traditional sense. In recent Food Marketing Institute (FMI) surveys, more than 30% of the husbands of women who work full-time did as much cooking, cleaning, and food shopping as their wives (Davis, 1989; FMI, 1988).

The population is aging. Going into the next century, the fastest growing groups will be those aged 45 to 54, along with those over age 85. By the year 2000, the proportion of Americans over age 65 will be the same throughout the country as the proportion in Florida today. An aging population means decreased discretionary spending and more demands for healthful and nutritious foods.

Minorities are growing in America. Within 10 years, one-quarter of all Americans will be either black, Hispanic, or Asian. The city of Los Angeles illustrates the trend. At present, Los Angeles is the largest Mexican city outside Mexico, the second largest Chinese city outside China, the second largest Japanese city outside Japan, and the largest Philippine city outside the Philippines (Davis, 1989).

The consumer demand for convenience, gourmet foods, ethnic items, and other services is increasingly evident in the food service and retail food industries. As the number of working women and single dwellers increases, the consumer base continues to change. With reduced leisure time, consumers who once spent two hours per day in the kitchen now spend less than a half hour. Convenience stores, fast-food restaurants, specialty food service outlets, and prepared items in the supermarket are food industry responses (FMI, 1988; Taylor, 1989).

To illustrate the impact of less preparation time in the home, a quick review of consumer buying habits is in order. In 1973, almost 80% of the food dollar was spent on home-prepared foods. In 1988, this number had fallen to 67%. Many predict that the figure may be as low as 40% by the year 2000. As with all foods, fish and shellfish preparation must be viewed in the manner in which consumers use the product in a contemporary environment.

This does not mean that the consumer will be eating at home less but, rather, that less time will be devoted to food preparation. This trend toward "cocooning," in which the family spends more time around the home but utilizes the time more prudently, is central to future consumer patterns (Davis, 1989).

Consumers want more convenience and nutrition. Value-added products, ready-to-eat items, and microwave entrees are examples. Deli departments of the supermarket may soon become food service operations, competing with fast-food and takeout restaurants (FMI, 1989; Taylor, 1989).

Seafood, like other foods, will be placed in a competitive consumer environment. Fish and shellfish must continue to taste good if they are expected to attract more consumers. Further, seafood must stay within the budget of the new consumer. If the industry can respond to the changing consumer base, the opportunity to expand per capita consumption appears good (Taylor, 1989).

The amount of imported product is not yet recognized as a potential problem by the consumer, yet it is of significant concern to regulatory officials. Rising needs place increased pressure on government to protect consumers without the ability to monitor the harvest, processing, and distribution of the hundreds of species in question. Because of the potential of ever-increasing imports, the safety issue is becoming a matter of international concern. Although agencies routinely sample and require country-of-origin labeling, the consumer is unaware of the complexity of attempting to truly safeguard these foodstuffs.

ACTIVITIES IN OTHER COUNTRIES

A number of countries have endeavored to enhance the value of their seafood products by enacting programs to ensure product quality. Canada, Denmark, and Norway have given high priority to marketing safe, quality seafood items. Canada, for example, inspects vessels, landing sites, and processing facilities on an annual basis. Vessels must meet the same exacting standards as processing facilities or risk losing their certification. Canadian plant registration requires compliance with a posted list of standards. At inspection, plants are rated by use of a Hazard Analysis Critical Control Point (HAACP) approach. Critical findings result in more frequent inspections or the possibility of noncertification.

In Europe, similar programs are in place. Denmark inspects fishing vessels. Each participant must meet certain sanitation requirements, as well as certification for activities such as on-board processing. Distribution centers receive regular inspections that monitor all products entering the marketplace. The advent of the European Economic Community (EEC) has brought forth a host of new regulations, ensuring that member nations comply with the policies of their EEC partners.

Many other countries have seafood inspection programs, but they are often not dedicated programs like those in Canada, Denmark, Norway, Iceland, and New Zealand. Consequently, they do not pay the same rigorous attention to detail. Indeed, most countries have programs centered on seafood as a food group, not as a distinct entity that requires special attention.

CONCLUSIONS AND RECOMMENDATIONS

Based on commercial sources, Americans consumed almost 23% more seafood in 1989 than they did 10 years earlier. This increase in consumption was not accompanied by a concomitant increase in reported seafood-borne illnesses. The total supply of fishery products to fulfill the domestic requirement for seafood was in excess of 8.5 billion pounds in 1989, with over 300 species involved in the catch statistics. Production and consumption trends suggest that domestic seafood demand will continue, with more emphasis on prepared convenience foods along with the traditional demand for fresh and frozen selections. Production will have to be supplemented with more imported and cultured sources. Recreational harvesting, both in the purist sense and as subsistence fishing, continues to contribute a significant portion to the annual per capita intake.

The committee recommends the following:

- Consumer information studies must be conducted to monitor the rapidly changing consumption trends in the United States. Patterns of consumer use and preparation, as well as sources of seafood products used in the home, must be evaluated. By better understanding consumption patterns, fishery managers and food regulators will be more able to influence dietary intake, and reduce potential exposure to fish from contaminated water.

- Changes in consumption patterns necessitate more attention to informing consumers on how to best handle highly perishable products such as seafood. As much as 50% of all reported, acute fish and shellfish problems might be eliminated by more careful handling and proper preparation in the home or in food service establishments. With the advent of more prepared foods, every effort should be made to ensure the safety of the product both in the manufacturing/distribution chain and for the end user.

- The retail and institutional handling of seafood products requires increased attention to control cross-contamination. A number of seafood-related illnesses can be traced to poor sanitation practices by employees or to lack of proper handling via the distribution system. More efforts will be needed to alert all users to the importance of time/temperature relationships, HACCP concepts, good manufacturing practices, and new technology (e.g., live holding tanks).

- Aquaculture promises to produce a larger share of domestically consumed fish and shellfish in the years ahead. Cultured plants and animals hold the promise of being high quality, and generally free of some of the contamination associated with wild species. Care, however, must be taken to avoid the untimely use of antibiotics and other chemicals in these closed or recirculated systems, which are often used to control pathogens in semiclosed systems.

- The safety of recreationally harvested fish and shellfish requires increased vigilance, which means increased focus on the origin, handling, and distribution of recreational products. These harvesting efforts may now account for over 20% of all fresh and frozen seafood consumed in the United States. However, this catch is not well controlled, and users may handle, distribute, and prepare the product in an unsafe manner. Further, much of this product may be harvested from areas not suited for consumption due to natural or induced contamination problems. Increased educational activity is required to protect the consumer with regard to this resource. Fishery managers will have to pay greater attention to the implications of sport caught fish and shellfish on consumer health.

7

Microbiological and Parasitic Exposure

Seafood, like any food item, has the potential to cause diseases from viral, bacterial, and parasitic pathogens under certain circumstances. These agents are acquired from three sources:

(a) mainly fecal pollution of the aquatic environment,

(b) to a lesser extent, the natural aquatic environment, and

(c) industry, retail, restaurant, or home processing and preparation.

With the exception of foods consumed raw, however, the reported incidences of seafood-related disease are low. Available data from the Centers for Disease Control and the Northeast Technical Support Unit of the Food and Drug Administration for 1978-1987, as well as literature reports, suggest that the greatest numbers of seafood-associated illnesses are from raw molluscan shellfish harvested from waters contaminated with raw or poorly treated human sewage.

The vast majority of this illness is gastroenteritis of unknown etiologies clinically suggestive of human-specific Norwalk and Norwalk-like agents. Although these are the most common seafood-associated illnesses, they tend to be relatively mild with no associated mortality.

Naturally occurring marine *Vibrio* species are responsible for many fewer reported cases of infection from the consumption of raw molluscan shellfish, but certain species such as *V. vulnificus* can be associated with high mortality (<50%) in persons who are immunocompromised or have underlying liver disease. The microbiological risk associated with seafood other than raw molluscan shellfish is much lower and appears to result from recontamination or cross-contamination of cooked by raw product, or from contamination during preparation followed by time/temperature abuse. This occurs mainly at the food service (postprocessing) level, which is common to all foods and not specific for seafood products.

Seafood-related parasitic infections are even less common than bacterial and viral infections, with *Anisakis simplex* and cestodes having the greatest public health significance in the United States. In general, parasitic infections are concentrated in certain ethnic groups that favor consumption of raw or partially cooked seafoods.

Thorough cooking of seafood products would virtually eliminate all microbial and parasitic pathogens; it will not destroy some microbial toxic metabolites (e.g., *Staphylococcus* toxins). Individuals who choose to eat raw seafood should be educated about the potential risks involved and how to avoid or mitigate them. In particular, immunocompromised individuals, those with defective liver function, people afflicted with diabetes, and the elderly should be warned never to eat raw shellfish.

The greatest risks from the consumption of raw molluscan shellfish could be minimized by research to develop valid indicators of human enteric viruses for proper classification of shellfish growing waters; by implementing and maintaining proper treatment and disposal of sewage to avoid human enteric pathogen contamination of harvest areas; by efforts to identify and limit the number of pathogenic *Vibrio* species in shellfish; by the development of new diagnostic methods and improved processing technology; and by the application of risk-based regulatory control measures for potential microbial pathogens in raw molluscan shellfish. Other seafood-associated risks can be reduced by proper application of a Hazard Analysis Critical Control Point system. This cannot be achieved by the visual or organoleptic inspection currently used for meat and poultry. Seafood inspection requires the development of valid microbiological guidelines to accurately assess human health risk from raw and processed seafoods. Inspection system guidelines must apply to imported as well as domestic products.

INTRODUCTION

Like any food items, fish and shellfish carry a variety of bacteria, viruses, and parasites capable of causing disease in consumers (WHO, 1990). Table below lists some of the agents that occur naturally in seafood or in the marine environment, are associated with sewage contamination of harvesting areas, or can be acquired during seafood harvest or processing. Many of these microorganisms pose only a slight risk to normal human populations, but all are pathogens and some pose serious risk to specific population groups, such as persons with defects in their immune systems (Archer and Young, 1988). Because of the increasing availability of sophisticated microbiological techniques, it has become possible to identify and provide detailed characterizations of many of the microorganisms present in or on seafood. Unfortunately, epidemiological studies, which are necessary to define risk clearly, have not kept pace with microbiological advances. In many instances, only rudimentary epidemiological data are available with which to correlate the information derived from microbiological product analyses. Thus, it is very difficult to assess the risk from these microorganisms to the health of the population.

The major sources of information on seafood-associated illness are the Centers for Disease Control (CDC) Foodborne Disease Outbreak Surveillance Program and a data base on shellfish-associated food-borne cases maintained by the Food and Drug Administration (FDA) Northeast Technical Support Unit. The CDC data are derived from reports of food-borne outbreaks[1] by state health departments. Reporting is passive, but data are collected in a systematic fashion. The FDA Northeast Technical Support Unit (NETSU) data come from books, news accounts, CDC reports, city and state health department files, Public Health Service regional files, case histories, and archival reports. Both collection systems have a number of inherent biases, which are discussed elsewhere in this report.

TABLE: Seafood-Associated Human Pathogens

Pathogens	Isolated from Seafoods	Proven Pathogen in Seafood	Pathogen Source[a]
Organisms That Can Cause Disease in Normal, Healthy Adults			
Bacteria			
Vibrio cholerae O1	Yes	Yes	1, 2
Vibrio cholerae non-O1[b]	Yes	Yes	1
Vibrio parahaemolyticus	Yes	Yes	1
Vibrio mimicus	Yes	Yes	1
Vibrio fluvialis	Yes	Yes	1
Vibrio furnissii	Yes	Yes	1
Vibrio hollisae	Yes	Yes	1
Salmonella typhi[c]	Yes	Yes	2, 3
Salmonella (nontyphoidal)	Yes	Yes	2, 3
Campylobacter jejuni	Yes	Yes	2, 3
Escherichia coli	Yes	No	2, 3
Yersinia enterocolitica	Yes	No	2, 3
Clostridium botulinum	Yes	Yes	2, 3
Shigella	Yes	Yes	2, 3
Staphylococcus aureus	Yes	Yes	3
Helminths			
Anisakis simplex	Yes	Yes	1
Other helminths	Yes	Yes	1
Viruses			
Poliovirus	Yes	No	2
Other picornaviruses	Yes	No	2
Norwalk/Snow Mountain/small round viruses (SRVs)	No	Yes	2
Enteral non-A, non-B hepatitis	No	Yes	2
Hepatitis A	Yes	Yes	2, 3
Organisms That Cause Disease Most Often in Special Population Gruops			
Vibrio vulnificus[d]	Yes	Yes	1
Rotavirus[e]	Yes	No	2
Listeria	Yes	No	1, 3

Organisms with Uncertain Roles as Food-borne Pathogens			
Aeromonas hydrophia[f]	Yes	Yes	1
Plesiomonas shigelloides	Yes	Yes	1
Edwardsiella tarda	Yes	No	1

[a] (1) Harvest water/associated with naturally occurring aquatic bacteria; (2) harvest water/associated with fecal pollution; (3) associated with processing and preparation (cross-contamination or time/temperature abuse, infected food handlers).

[b] Causes gastroenteritis in normal, healthy hosts; can cause septicemia in persons in high-risk groups, as outlined in Table 3-6.

[c] Primarily of historical association in the United States, but remains a problem in some foreign countries and could affect imports.

[d] Illness usually confined to high-risk groups outlined in Table 3-6.

[e] Illness generally occurs in children under the age of 2; older persons are usually immune.

[f] *Aeromonas* can cause serious wound infections and septicemia; however, conclusive data on its role as a cause of gastroenteritis are lacking. Studies suggesting that it is a gastrointestinal pathogen have not implicated seafood as a risk factor for illness.

Based on experience with other foods (NRC, 1987), it is likely that only a small fraction of seafood-associated disease is reported and that the two available data bases therefore reflect only a small fraction of the actual number of seafood-associated illnesses that occur. Even when outbreaks are reported, etiologic agents are frequently not identified. For example, the NETSU data base from 1978 to 1987 includes an additional 5,342 cases of shellfish-associated illness for which no etiology was determined; the CDC food-borne surveillance data base reported 3,271 cases of shellfish-associated illness and 203 cases of other seafood-associated illness with unknown etiologies in the same time period. Cases with unknown etiology are probably not all of microbiological origin and could include toxins or allergies, among other causes. Overall, because of the different surveillance and reporting systems, the two data bases do not consistently correlate reports of outbreaks and cases of the same pathogens.

TABLE : Seafood-Associated Outbreaks and Related Cases by Pathogen, 1978-1987

	Finfish/Other		Shellfish[a]		
	CDC		CDC		NETSU
Pathogen	Outbreaks	Cases	Outbreaks	Cases	Cases
Naturally Occurring Aquatic Agents					
Vibrio parahaemolyticus[b]	-	-	-	-	52
Vibrio cholerae O1	1	2	2	14	13
Vibrio cholerae non-O1	-	-	2	11	120
Vibrio vulnificus	-	-	-	-	100
Vibrio mimicus	-	-	-	-	5
Vibrio hollisae	-	-	-	-	5
Vibrio fluvialis	-	-	-	-	5
Plesiomonas	-	-	-	-	18
Aeromonas	-	-	-	-	7
Giardia	1	29	-	-	-
Diphyllobothrium	1	10	-	-	-
Infectious Agents Generally Associated with Fecal Pollution					
Unspecified hepatitis[b]	-	-	-	-	1,645
Hepatitis A[b]	-	-	7	33	45
Salmonella (nontyphoidal)[b]	-	-	3	80	-
Shigella	-	-	4	77	84
Campylobacter	-	-	-	-	16
Norwalk and related viruses	-	-	2	42	82
Non-A, non-B hepatitis[b]	-	-	-	-	1
Infectious Agents Generally Associated with Processing and Preparation					
Vibrio parahaemolyticus[b]			15	176	-
Clostridium perfringens	1	46	2	28	-
Hepatitis type A[b]	2	92	-	-	-
Salmonella (nontyphoidal)[b]	3	67	-	-	-
Clostridium botulinum	26	38	-	-	-
Shigella[b]	3	60	-	-	-
Staphylococcus aureus	1	3	1	9	-
Bacillus cereus	1	4	2	6	-
Total outbreaks/cases	38	351	40	476	2,198
Unknown agents[c]	16	203	88	3,271	5,098

TABLE : Shellfish-Associated Outbreaks and Related Cases by Frequency of Occurrence, 1978-1987

CDC				NETSU Cases	
Outbreaks		Cases			
Etiology	Number	Etiology	Number	Etiology	Number
1. Unknown[a]	88	1. Unknown[a]	3,271	1. Unknown[a]	5,098
2. *Vibrio parahaemolyticus*	15	2. *Vibrio parahaemolyticus*	176	2. Unspecified hepatitis	1,645
3. Hepatitis A virus	7	3. *Salmonella* (non-typhi)	80	3. Non-O1 *Vibrio cholerae*	120
4. *Shigella*	4	4. *Shigella*	77	4. *Vibrio vulnificus*	100
5. *Salmonella*	3	5. Other viral	42	5. *Shigella*	84
6. *Vibrio cholerae* O1	2	6. Hepatitis A virus	33	6. Norwalk and related viruses	82
Other viral	2				
Non-O1 *Vibrio cholerae*	2	7. *Clostridium perfringens*	28	7. *Vibrio parahaemolyticus*	52
Clostridium perfringens	2				
Bacillus cereus	2	8. *Vibrio cholerae* O1	14	8. Hepatitis A virus	45
7. *Staphylococcus aureus*	1	9. Non-O1 *Vibrio cholerae*	11	9. *Plesiomonas*	18
		10. *Staphylococcus aureus*	9	10. *Campylobacter*	16
		11. *Bacillus cereus*	6	11. *Vibrio cholerae* O1	13
				12. *Aeromonas*	7
				13. *Vibrio mimicus*	5
				Vibrio hollisae	5
				Vibrio fluvialis	5
				14. Non-A, non-B hepatitis	1
Totals	128		3,747		7,280

[a] The unknown etiologies are probably not all microbiological pathogens.

SOURCE: CDC (1989); Rippey and Verber (1988).

Despite these limitations, they represent the only available national data bases on finfish- and shellfish-associated diseases. In this chapter, these data bases are used as a starting point to assess the relative importance of seafood-associated pathogens and to evaluate, to the extent possible, the risk that each pathogen poses to consumers. Risk management is dictated to a large degree by where and how microorganisms contaminate seafood or where they may be most easily controlled. For this reason, pathogens have been grouped according to their origin. The natural marine or freshwater environment harbors specific bacterial and helminthic parasitic pathogens, whereas pollution contributes bacterial and viral pathogens from human and animal fecal sources.

Microbial agents associated with workers or the environment of processing, distribution, and food service systems include both anthropophilic microorganisms and microorganisms that populate reservoirs of infection created by processing conditions.

TABLE : Finfish and Other Seafood-Associated Outbreaks and Related Cases by Frequency of Occurrence, 1978-1987

Pathogen	Outbreaks	Pathogen	Cases
1. *Clostridium botulinum*	26	1. Unknown[a]	203
2. Unknown[a]	16	2. Hepatitis A virus[b]	92[a]
3. *Salmonella* (nontyphoidal)	3	3. *Salmonella*	67
Shigella	3		
		4. *Shigella*	60
4. Hepatitis A virus[b]	2[a]		
		5. *Clostridium perfringens*	46
5. *Staphylococcus aureus*	1		
Clostridium perfringens	1	6. *Clostridium botulinum*	38
Vibrio cholerae O1	1		
Bacillus cereus	1	7. *Bacillus cereus*	4
		8. *Staphylococcus aureus*	3
		9. *Vibrio cholerae* O1	2
Total outbreaks/cases	51		515

[a] Unknown etiologies are probably not all microbiological pathogens.

[b] Food handler positive for hepatitis A virus.

SOURCE: CDC (1981b,c, 1989).

This chapter emphasizes domestic production of wild caught fish and shellfish. The same pathogens are of concern in imported seafood, although the risks of specific pathogens vary depending on conditions in the growing waters at the point of harvest, as well as subsequent handling and processing. Aquaculture presents a different set of potential concerns, which are summarized in a separate section below.

PATHOGENS NATURALLY PRESENT IN MARINE OR FRESHWATER ENVIRONMENTS

Naturally Occurring Marine Bacteria Associated with Human Disease

A number of free-living estuarine and freshwater bacteria may be associated with human disease. Most of these bacteria fall within the family Vibrionaceae, which includes the genera *Vibrio, Aeromonas* , and *Plesiomonas*.

TABLE: *Vibrionaceae Identified with Shellfish-Associated Human Disease*

Pathogen	Clinical Features		Implicated Shellfish Vehicle	
	Diarrhea	Septicemia	Most Common	Other
Vibrio cholerae O1	++[a]		Crab	Oyster
Vibrio cholerae non-O1	++	+[b]	Oyster	
Vibrio parahaemolyticus	++	(+)[c]	Shellfish	
			(shrimp, crab)	
Vibrio fluvialis	++		?[e]	
Vibrio mimicus	++		Oyster	
Vibrio hollisae	++	(+)	Shellfish	
			(unspecified)	
Vibrio furnissii	++		?	
Vibrio vulnificus	+	++	Oyster	Crab
Aeromonas hydrophila	?[d]		?	
(including *Aeromonas caviae* and *Aeromonas sobria*)				
Plesiomonas shigelloides	++		Oyster	

[a] Most common clinical manifestation.
[b] Occasional clinical manifestation.
[c] Rare clinical manifestation.
[d] Suspected, but not proven clinical manifestation.
[e] Unsure if shellfish can be implicated as a vehicle.

These bacteria are generally not associated with fecal contamination of harvest waters, and some studies suggest an inverse relationship between counts of certain species and fecal coliform levels (Kaper et al., 1979; Tamplin et al., 1982). Counts tend to be highest in warm summer months, particularly when water temperature exceeds 15-20°C (Baross and Liston, 1970).

Illnesses associated with these organisms can generally be divided into two categories: disease (usually gastroenteritis) due to ingestion of seafoods containing these organisms, and wound infections related to contamination of wounds by seawater (Blake et al., 1979; Morris and Black, 1985). Cases tend to occur during late summer and early fall, when bacterial counts are highest in the water.

***Vibrio cholerae* O1.**

Vibrio cholerae can be classified according to O group: strains in O group 1 (*V. cholerae* O1) cause cholera, whereas strains in other O groups (non-O1 *V. cholerae*) are generally associated with milder illness. The virulence of *V. cholerae* O1 is determined primarily by the presence of a protein enterotoxin, cholera toxin (CT). Strains that do not produce cholera toxin (i.e., are not toxigenic) tend to be avirulent or have reduced virulence (Morris et al., 1984). Persons with the most severe forms of cholera (cholera gravis) have profuse, watery diarrhea (Pierce and Mondal, 1974). The volume of diarrhea may exceed 1 liter per hour, resulting in rapid water depletion, circulatory collapse, and (if untreated) death. Fortunately, cholera gravis is relatively uncommon. In infections with *V. cholerae* O1 of the classical biotype, four or five inapparent infections or mild illnesses may occur for every apparent case. In infections with El Tor (the biotype present in the United States), 25-100 other infections may be expected for every hospitalized case (Bart et al., 1970).

It has traditionally been thought that strains of *V. cholerae* O1 were transmitted by fecal contamination of food or water, and this mode of transmission likely predominates in developing countries. However, there is increasing evidence that free-living strains of *V. cholerae* O1 have become established in the U.S. Gulf Coast environment and may be transmitted to man via consumption of raw, undercooked, or cross-contaminated shellfish (Morris and Black, 1985). The number of cases of cholera associated with these strains is relatively small (in the range of 50 cases since 1973); their significance lies in their potential for causing severe disease in otherwise healthy hosts.

Epidemiology and Risk Assessment

Epidemiologic investigations have associated *V. cholerae* O1 illness with eating crabs, shrimp, and raw oysters harvested along the Gulf Coast (Blake et al., 1980). The CDC (1989) reported three outbreaks of *V. cholerae* O1 involving 16 cases between 1978 and 1987, and NETSU (Rippey and Verber, 1988) reported 13 cases in this same period. Based on these and other reports in the literature, at least 50 cholera cases appear to have been acquired in the United States since 1973 when the first recent indigenous U.S. case was reported (Blake et al., 1980; Lowry et al., 1989b; Morris and Black, 1985).

Toxigenic *V. cholerae* O1 has been isolated from estuarine water, from fresh shrimp, and from cooked crabs. All U.S. *V. cholerae* O1 isolates have been hemolytic, biotype El Tor, serotype Inaba, and all have had the same unique *Hin*dIII digest pattern on Southern blot analysis. Because of the occurrence of cases over a period of years and in a variety of locations, many authors suggest that this strain has become endemic along the Gulf Coast (Morris and Black, 1985).

Although toxigenic *V. cholerae* O1 strains are known to be present in Gulf Coast estuaries, the percentage of shellfish that carry the organism appears to be quite low. In one study conducted by the FDA, *V. cholerae* O1 was isolated from 0.9% of 790 oyster lots sampled over 12 months. None of the *V. cholerae* O1 strains isolated was able to produce cholera toxin (Twedt et al., 1981).

Seroepidemiologic studies provide an alternative means of assessing disease burden/risk of infection, particularly for microorganisms such as *V. cholerae* O1 that produce a high percentage of asymptomatic infections. In a study conducted along the Gulf Coast in Texas (Hunt et al., 1988), 0.89% of persons sampled had elevated titers of both vibriocidal and anticholera toxin antibodies, the standard serologic assays used for *V. cholerae* infections. These assays have relatively low specificity, making the interpretation of results difficult. However, these data do raise the possibility that a small percentage of persons living along the U.S. Gulf Coast has recently been infected with toxigenic strains of *V. cholerae* O1 (Hunt et al., 1988).

Disease Control

Because *V. cholerae* appears to contaminate marine animals in situ, it must be destroyed by treatment of the food. For crustacean seafood, proper cooking in primary processing (crab) or at the food service level (shrimp) and avoidance of recontamination of cooked product are recommended (Shultz et al., 1984). Studies conducted by CDC indicate that large 0 for less than 8 minutes or steamed for less than 25 minutes may still contain viable *V. cholerae* organisms (Blake et al., 1980), an observation that has led to a series of recommended time and temperature conditions for cooking crabs.

Toxigenic *V. cholerae* O1 strains can be identified rapidly in shellfish by using deoxyribonucleic acid (DNA) probes. A monitoring system that employs such probes might be useful in identifying potential "high-risk" harvesting areas. However, without further studies it is unclear how such data should be used. Given the low frequency with which toxigenic *V. cholerae* O1 are found, it would be difficult to justify embargoing all shellfish from a given area based solely on a positive sampling result. Using probes to screen shellfish on a lot-by-lot basis would be of little value, because results from a single crab or oyster are unlikely to be representative of the lot as a whole. Finally, epidemiological data are lacking to show how changes in the frequency of isolation of *V. cholerae* O1 from shellfish correlate with changes in disease occurrence in the community.

Non-O1 Vibrio cholerae (V. cholerae of O Groups Other Than 1)

Non-O1 *V. cholerae* strains are ubiquitous in estuarine environments (including bays and estuaries of the U.S. Gulf, Atlantic, and Pacific coasts) and are commonly isolated from shellfish. In one study conducted by the FDA, non-O1 *V. cholerae* was isolated in up to 37% of U.S. oyster lots harvested during warm summer months (Twedt et al., 1981). Non-O1 *V. cholerae* has been associated with gastroenteritis, wound and ear infections, and septicemia (Hughes et al., 1978; Safrin et al., 1988). Gastroenteritis can occur in normal, healthy persons (Morris et al., 1990). Septicemia appears to occur primarily in persons who are immunocompromised or who have underlying liver disease; the mortality rate for persons with septicemia exceeds 50% (Safrin et al., 1988). Persons who have acquired non-O1 *V. cholerae* infections in the United States have almost all given a history of having eaten raw oysters before the onset of illness (Morris et al., 1981). However, the number of reported cases of non-O1 *V. cholerae* gastroenteritis is relatively low, suggesting that only a minority of strains are able to infect humans, or that most infections result in mild or asymptomatic illness.

Epidemiology and Risk Assessment

In the NETSU data base, non-O1 *V. cholerae* is the most common bacterial cause of molluscan shellfish-associated illness. Still, only 120 cases were reported between 1978 and 1987 (Rippey and Verber, 1988). The CDC (1989) reported only two shellfish-associated outbreaks involving 11 cases during the same period. Given the amount of raw shellfish consumed and the frequency with which the organism is present in shellfish, the number of reported cases of non-O1 disease in the United States is much less than might be anticipated. Seven coastal area hospitals in four southern states isolated only seven specimens of non-O1 *V. cholerae* (including five isolates from a single outbreak) from approximately 11,000 stool cultures performed with the use of thiosulfate-citrate-bile salts-sucrose (TCBS) agar, an appropriate selective culture medium (Morris and Black, 1985). Only two non-O1 *V. cholerae* specimens were isolated from over 10,000 stool cultures on TCBS performed during 14 years in a Chesapeake Bay area hospital (Hoge et al., 1989).

This small number of reported cases is probably a reflection of several factors. It is likely that the majority of environmental strains lack the necessary colonization factors, appropriate toxins, or other virulence determinants to cause human disease

(Morris et al., 1990). Even when infections occur, patients may often be asymptomatic or have only mild illness and, consequently, not come to medical attention. In support of the latter hypothesis, non-O1 *V. cholerae* strains were isolated from 13 (2.7%) of 479 persons in a cohort of physicians attending a convention in New Orleans in late September; among persons eating raw oysters, 4% had non-O1 *V. cholerae* in their stool. However, despite this relatively high colonization rate, only 2 (15%) of the 13 culture-positive persons were symptomatic, comparable to the overall 14% rate of diarrhea reported in the entire cohort (Lowry et al., 1989a).

Disease Control

There are currently no programs attempting to limit exposure of the general population to non-O1 *V. cholerae*, other than recommendations that persons who are immunocompromised or who have underlying liver disease avoid the consumption of raw oysters (Blake, 1983). Exposure to non-O1 strains could be reduced significantly if oyster harvesting was confined to colder months when *Vibrio* counts in water are lowest (Twedt et al., 1981).

If it is assumed that most environmental non-O1 *V. cholerae* strains are nonpathogenic, identification of these strains in oysters has limited public health utility. It would clearly be of value if pathogenic strains could be differentiated from those that are nonpathogenic. Basic research in this area should be encouraged.

Vibrio Parahaemolyticus

The CDC reported *V. parahaemolyticus* as the most common cause of vibrio disease due to consumption of seafoods from 1978 to 1987, whereas NETSU reported lower incidences. This is a mildly halophilic vibrio commonly isolated from fish, shellfish, and other marine sources in inshore waters, which is most abundant when water temperatures exceed 15°C. It is difficult to isolate during cold winter months. The ability to cause human gastroenteritis is most highly correlated with the production of a heat-stable hemolysin (Miyamoto et al., 1969). Most strains isolated from the marine environment lack this hemolysin and are probably not pathogenic, although nonhemolytic strains have recently been associated with illness occurring along the U.S. Pacific Coast (Abbott et al., 1989; Kelly and Stroh, 1989). *V. parahaemolyticus* reproduces very rapidly at temperatures of 20°C and above, and has been shown to reach potentially infective levels [more than 10^5 colony forming units (CFU)] in shrimp and crabs held for 2-3 hours at such temperatures (Liston, 1973). However, it is heat sensitive and rapidly killed at 60°C.

Epidemiology and Risk Assessment

V. parahaemolyticus is a common marine isolate, with isolation reported from water, sediment, suspended particulates, plankton, fish, and shellfish (Joseph et al., 1983). However, it is likely that only a small fraction of marine isolates are potentially pathogenic. For example, in a study by Thompson and Vanderzant (1976), only 4 of 2,218 isolates from Galveston Bay were able to produce the heat-stable hemolysin generally associated with virulence.

In Japan, *V. parahaemolyticus* has been implicated as the etiologic agent in 24% of reported cases of food-borne disease (Miwatani and Takeda, 1976). In the United States, *V. parahaemolyticus* has caused several major food-borne disease outbreaks (Barker, 1974). The CDC food-borne surveillance data from 1978 to 1987 reported *V. parahaemolyticus* as the cause of 15 outbreaks associated with 176 cases of mostly crustacean shellfish-associated illness (CDC, 1989). No other seafood-associated illnesses from *V. parahaemolyticus* were reported to CDC during this time; NETSU reported 52 cases in the same period (Rippey and Verber, 1988).

Outbreaks of *V. parahaemolyticus* have often been associated with cross-contamination or time/temperature abuse of cooked seafood. Although sporadic cases associated with the consumption of raw oysters have occurred, there does not appear to be the same strong association with consumption of raw oysters as reported for non-O1 *V. cholerae*. As with non-O1 *V. cholerae*, estimates of incidence are complicated by the need for TCBS selective media for isolation of the organism from stool cultures. In areas and studies in the United States where TCBS has been used routinely, *V. parahaemolyticus* has been the *Vibrio* species isolated most frequently from stool samples, with an isolation rate two to three times that reported for non-O1 *V. cholerae* (Bonner et al., 1983; Hoge et al., 1989; Lowry et al., 1989a). However, the overall disease incidence remains quite low, even in communities where shellfish consumption would be expected to be high. In a hospital-based study in Annapolis, Maryland, there was estimated to be less than 0.5 case/100,000 population/year (Hoge et al., 1989).

Disease Control

The impact of major food-borne outbreaks due to *V. parahaemolyticus* can be limited by emphasizing avoidance of cross-contamination and time/temperature abuse of cooked shellfish or of seafood that will be consumed raw. Illness due to *V. parahaemolyticus* is typically associated with multiplication of the organism to high numbers on the food either in the raw state or after cooking. Seafood eaten shortly after adequate cooking is generally safe (Bradshaw et al., 1974). Control has thus been aimed at ensuring good handling and processing practices whereby the seafood is held at low temperature (<5°C) and recontamination of the cooked product is avoided.

Specific counts of *V. parahaemolyticus* on seafoods have been proposed as an index of hazard, but this method has not been widely used because of cost, lack of specificity for pathogenic strains, and time lag (3-5 days) between sampling and results.

The DNA probes for identification of hemolytic (Kanagawa-positive) strains are available (Nishibuchi et al., 1985). With nonradioactively labeled oligonucleotide probes it may be possible to identify potentially pathogenic strains within a matter of hours. However, no effort has been made to make such probes commercially available or to assess their utility as public health tools.

Vibrio Vulnificus

V. vulnificus is very similar to *V. parahaemolyticus* in cultural characteristics and sensitivity to processing procedures. It is widespread in estuarine waters around the U.S. coastline and is a common isolate in oysters harvested from warm (>20°C) waters. Although *V. cholerae* O1 is capable of causing severe, life-threatening disease in healthy individuals, *V. vulnificus* is potentially the most dangerous of the vibrios for persons with underlying illnesses such as cirrhosis or hemochromatosis or persons who are immunocompromised (Klontz et al., 1988; Morris et al., 1987b; Tacket et al., 1984). When such persons become infected, the mortality rate can exceed 50% (Morris, 1988). Fortunately, *V. vulnificus* infections are relatively uncommon, with an estimated annual incidence rate in the range of 0.5 case/100,000 population/year.

Epidemiology and Risk Assessment

Like other *Vibrio* species, *V. vulnificus* appears to be part of the normal bacterial flora of estuaries along the U.S. Gulf, Atlantic, and Pacific coasts. Although data are limited, some studies reported that close to 100% of oyster lots harvested in some areas during warm summer months have been positive for *V. vulnificus*, and in one study, 11% of the blue crabs harvested from Galveston Bay during summer months had the organism in their hemolymph (Davis and Sizemore, 1982; Tamplin, 1990).

TABLE : Risk Factors for Primary Vibrio Septicemia

Risk Factors[a]	CDC (n = 42)[b] %	Florida (n = 38)[c] %
Alcoholic cirrhosis	31	32
Hemochromatosis/cirrhosis	12	13
"Chronic hepatitis"	10	-
Postnecrotic cirrhosis	5	-
Liver disease-undetermined or unspecified cause[d]	10	21
Alcohol abuse (without liver disease)	12	16
Diabetes	10	11
Other illnesses[e]	3	3
No known underlying disease	7	5

[a] Based on report of attending physician.

[b] Includes data from national retrospective studies conducted by Blake et al. (1979) and Tacket et al. (1984) at the Centers for Disease Control.

[c] Data from a prospective study conducted in Florida by Klontz et al. (1988).

[d] May include patients with alcohol-related illness.

[e] These include renal disease, steroid-dependent rheumatoid arthritis, thalassemia, malignancies, hematologic disorders associated with immunosuppression (neutropenia and leukopenia), and acquired immune deficiency syndrome.

Occurrence of *V. vulnificus*-induced primary septicemia is significantly associated with eating raw oysters (Tacket et al., 1984). However, susceptibility appears to be limited to certain high-risk groups, particularly those with hemochromatosis, cirrhosis, hematologic or other disorders associated with immunosuppression, renal failure, and diabetes. Wound infections are associated with contamination of wounds with seawater. Although wound infections can occur in otherwise healthy persons, deaths associated with wound infection have occurred almost exclusively among persons in the same high-risk groups mentioned above.

Between 1978 and 1987, NETSU reported 100 cases of *V. vulnificus* infections (Rippey and Verber, 1988). *V. vulnificus* infections have not been reported to CDC because outbreaks involving two or more individuals have not occurred. The increased number of cases reported to NETSU in the last two years probably reflects better isolation methods and a more rigorous screening for the organism by public health laboratories. However, the mortality rate has remained high.

TABLE:* Vibrio vulnificus *Cases Associated with Oysters, 1978-1987

	Vibrio vulnificus Cases	Percentage of Total Oyster-Associated Cases	Deaths	Mortality (%)
1978	2	4	0	0
1979	7	14	2(?)	29
1980	2	2	1	50
1981	16	64	5	31
1982	13	2	10	77
1983	9	12	4	44
1984	9	6	4	44
1985	7	16	0	0
1986	21	14	5	23
1987	16	68	8	53

SOURCE: NETSU data base (Rippey and Verber, 1988).

The incidence of *V. vulnificus* infections in coastal states appears to be 0.5 case/100,000 population/year, with primary septicemia accounting for approximately two-thirds of the cases.

Based on reported cases, the incidence in the Florida panhandle was estimated to be 0.4 case/100,000/year (Klontz et al., 1988). Johnston et al. (1985) estimated an incidence of 0.8/100,000/ year in Louisiana coastal parishes. The estimated incidence in a hospital-based study in Annapolis, Maryland was 0.5 case/ 100,000/year (Hoge et al., 1989). These estimates are clearly lower than might be expected, given the frequency with which the organism has been identified in shellfish and the amount of shellfish consumed in these areas. This may reflect differences in virulence among strains, the effect of infectious dose (i.e., the minimum number of organisms needed to cause disease), or the critical importance of host susceptibility in the disease process.

Disease Control

The incidence of *V. vulnificus* infections associated with ingestion of raw oysters by high-risk individuals could be reduced by cooking the product before eating. If a dose-response relationship exists, disease incidence might also be reduced by reducing the counts of *V. vulnificus* in shellfish by restricting harvesting to times when vibrio counts in harvesting areas are low (i.e., when water temperatures are <20°C) or by monitoring vibrio counts carefully in summer months. A recent study by Cook and Ruple (1989) indicated increases of one to three orders of magnitude in subsequent storage of shellstock at above 20°C. Rapid cooling of shellfish after harvest may decrease the initial bacterial counts, and careful avoidance of time/ temperature abuse during handling and shipment would limit subsequent multiplication of the organism (Cook and Ruple, 1989).

Low-dose gamma irradiation of live shellstock and fresh or frozen seafood products has been shown to be extremely effective in the elimination of *Vibrio* species (Giddings, 1984; Kilgen et al., 1988). Rapid diagnostic techniques are available for *V. vulnificus*, based both on agglutination reactions with specific antisera (Simonson and Siebeling, 1986) and on DNA probes (Morris et al., 1987a). However, these techniques have not been used as regulatory tools, and their exact role in disease control remains to be determined.

An ability to differentiate potentially pathogenic strains from strains without recognized virulence would clearly be of help in designing appropriate control strategies (Simpson et al., 1987). Similarly, there is a need to better understand the interaction between the organisms and oysters, both before and after harvesting. A better understanding of the host response to *V. vulnificus* would allow more careful delineation of groups at high risk for infection. An educational program defining the risks and explaining prevention by cooking and proper handling should be directed to this group through physicians, health organizations, and other public sources. Education of seafood producers, seafood processors and handlers, health professionals, and consumers regarding groups at high risk is also extremely important. Further studies in these areas must be encouraged as part of any overall plan for control of this pathogen.

Other Vibrio Species

Several other *Vibrio* species have been associated with human illness, including *V. mimicus* (Shandera et al., 1983), *V. hollisae* (Morris et al., 1982), *V. fluvialis* (Huq et al., 1980), and *V. furnissii* (Brenner et al., 1983). Illness from these species has generally been associated with seafood or shellfish consumption. However, the incidence of infection with these organisms appears to be quite low, and they are of much less importance as human pathogens than *V. cholerae, V. parahaemolyticus*, and *V. vulnificus*.

V. fluvialis, V. mimicus, and *V. furnissii* all appear to be naturally present in the marine environment. Infection with *V. mimicus* and *V. hollisae* has been associated with consumption of raw shellfish, especially raw oysters (Morris et al., 1982; Shandera et al., 1983). It is very difficult to estimate the incidence of infection with these organisms, because appropriate selective media for isolation are seldom used (except for *V. hollisae*, which may be best seen on blood agar) and because of difficulties with *Vibrio* taxonomy. Although they contain a number of inherent biases, data on the number of strains of each species sent to CDC for identification provide some basis for comparing relative frequencies of isolation. In a period during which CDC received 136 non-O1 *V. cholerae* isolates for identification, 39 *V. mimicus* isolates were received, 32 *V. hollisae* , 15 *V. fluvialis*, and 16 *V. furnissii* (Farmer et al., 1985). Between 1978 and 1987, NETSU (Rippey and Verber, 1988) reported five cases of *V. mimicus*, five cases of *V. hollisae*, and five cases of *V. fluvialis*.

The risk of infection with these organisms can likely be reduced by strategies similar to those used for other *Vibrio* species: that is, by using good manufacturing practices (GMPs) involving immediate and proper refrigeration, by possibly limiting shellfish harvesting to cooler months, by monitoring *Vibrio* counts in summer months for management purposes, by using low-dose irradiation for processing raw seafoods or live shellstock, and by advising persons in high-risk groups to avoid consumption of raw shellfish.

Aeromonas

Aeromonas species are common environmental isolates. They are ubiquitous in estuarine areas (Kaper et al., 1979) and can be isolated from a variety of foods (Palumbo et al., 1989). Their role as a pathogen remains highly controversial.

In controlled epidemiologic studies, strains within the *A. hydrophila* group have been associated with diarrheal disease in children in Australia and in travelers in Thailand (Burke et al., 1983; Pitarangsi et al., 1982). However, there are a number of studies in which no association with diarrheal disease has been established, and in which isolation rates from asymptomatic persons equal or exceed isolation rates from persons with diarrhea. In the one volunteer study performed, volunteers ingesting *Aeromonas* strains did not become ill. A small minority of strains may carry virulent characteristics that enable them to cause human disease, but there is currently no way of differentiating potentially pathogenic strains from nonpathogenic strains.

Contact with untreated surface water appears to be the major risk factor for infection with *Aeromonas* (Burke et al., 1983; Holmberg et al., 1986a). Between 1978 and 1987, only seven cases associated with shellfish consumption were reported to NETSU; CDC did not report any seafood-associated cases of

Aeromonas in this same period. Currently, any association between eating fish or shellfish and *Aeromonas* infection is at best circumstantial.

No programs to limit exposure to *Aeromonas* currently exist. Any such program would probably be premature because of uncertainties about the role of the organism as a pathogen.

Plesiomonas

Plesiomonas shigelloides (formerly *Aeromonas shigelloides*) has been implicated as a cause of human gastroenteritis for 40 years (Miller and Koburger, 1985). Like other members of the Vibrionaceae, *P. shigelloides* is a common environmental isolate, widespread in nature, being mostly associated with fresh surface water but also found in seawater (Van Damme and Vandepitte, 1980). It shows a seasonal variation in its isolation similar to that of marine vibrios, being isolated more often during warmer months (Miller and Koburger, 1985). Isolation of *P. shigelloides* has been significantly associated with foreign travel and with the consumption of raw oysters (Holmberg et al., 1986b; Kain and Kelly, 1989). Other foods implicated as vehicles for *P. shigelloides* include cuttlefish salad and salt mackerel. In the United States, raw oysters are probably the food most commonly implicated. However, questions remain about its pathogenicity.

Controlled epidemiologic studies have not shown a clear association between isolation of *P. shigelloides* and occurrence of diarrhea (Pitarangși et al., 1982), and volunteers ingesting the organism have not become ill (Herrington et al., 1987). It is possible that few strains carry virulent characteristics that enable them to cause human disease, but there is currently no way of differentiating potentially pathogenic strains from nonpathogenic ones.

According to the NETSU data base, *P. shigelloides* has been implicated in only 18 cases of illness in 1978-1987. This is less than 0.5% of the cases of illness associated with molluscan shellfish. The CDC did not report any outbreaks of seafood-associated *P. shigelloides* in the same period.

Most strains of *P. shigelloides* have a minimum growth temperature of 8°C, but at least one strain has been reported to grow at 0°C. Strains seem to survive well in shellstock oysters held at refrigeration temperature. The organism is sensitive to a pH less than 4 and to a salt concentration higher than 5% (Miller and Koburger, 1985). In addition, being a member of the family Vibrionaceae, it should be killed by relatively mild cooking temperatures.

Because of the strong association between eating oysters and human infection, *P. shigelloides* should be considered in any program to limit exposure to bacterial pathogens in shellfish. Again, there is a need to establish its pathogenicity for humans before initiating any specific control measures.

Helminthic Agents Present in Seafood

A number of helminths may be present in fish and shellfish. However, with few exceptions, they are harmless to humans. Illness from these agents is most commonly associated with eating fish. However, helminths are responsible for much less illness than either bacterial or viral agents. Only two outbreaks involving 39 cases of fish-associated parasitic illnesses were reported to CDC (1989) between 1978 and 1987, one outbreak each in California and Minnesota. Thus, parasitic infections from fish are rare in the United States, and there is, as yet, no evidence of a significant increase due to changed eating habits.

For many of these agents, humans are not the definitive host. Characteristics of the organism's life cycle may determine its relative importance as a human pathogen. Risk of exposure is dependent to a large degree on the geographic location of harvest. Areas endemic for specific helminth zoonoses in the United States include the Great Lakes region and Florida (*Diphyllobothrium latum*); Pacific Northwest and marine areas (*D. pacificum*); Pacific Islands, New Orleans, and Puerto Rico (*Angiostrongylus cantonensis*); the Atlantic and Pacific coasts (*Anisakis simplex*); the Pacific coast (*Contraceacum osculatum*); the Gulf of Mexico (*Contraceacum* spp.); the Northern Atlantic and Pacific Ocean (*Pseudoterranova dicipiens*); and the Pacific islands (*Paragonimus westermani*) (Bryan, 1986; Chitwood, 1970; Healy and Juranek, 1979).

TABLE: Helminthic Agents Present in Fish and Shellfish

Imported fish and shellfish
Trematodes
Clonorchis sinensis Heterophyes heterophyes Metagonimus yokogawai Opisthorchis felineus
Nematodes
Gnathostoma spinigerum
Domestic fish and shellfish
Trematodes
Paragonimus westermani
Cestodes
Diphyllobothrium latum Diphyllobothrium pacificum
Nematodes
Angiostrongylus cantonensis Contraceacum species (*osculatum*) *Hysterothylacium* species *Pseudoterranova dicipiens Anisakis simplex*

Trematodes

Only digenetic trematodes are known to produce disease in humans, and there are no data to suggest that any of these organisms represent a substantive health problem in the United States.

Clonorchis sinensis, or the Chinese liver fluke, is transmitted by eating raw or undercooked freshwater fish containing cyst stages of the organism. The organism is endemic in east Asia in a zone extending from Japan to Vietnam. *Heterophyes*, an intestinal fluke, is transmitted by eating fresh-or brackish-water fish (frequently mullet) in a raw, salted, or dried condition. The organism is a common parasite in the lower Nile valley near the Mediterranean coastline, and has been reported in the Orient and in western India. *Metagonimus yokogawai*, which closely resembles *H. heterophyes*, may be present in freshwater trout in the Orient, the maritime provinces of the USSR, northern Siberia, and the Balkans. *Paragonimus westermani*, the oriental lung fluke, is acquired by eating freshwater crabs or crawfish that are raw or have been pickled in brine, vinegar, or rice wine. The organism is endemic in east Asia, including Japan, Korea, the Philippines, and Taiwan.

Transmission of food-borne illness due to trematodes is prevented by adequate cooking and probably by freezing seafoods (Healy and Juranek, 1979).

Cestodes

Diphyllobothrium latum and *D. pacificum* are the only cestodes associated with parasitic infections from seafoods. However, there are inadequate data to estimate their incidence in the United States. Cases have been reported among Jewish cooks who sampled gefilte fish during its preparation.

D. latum is acquired by eating raw freshwater fish containing appropriate plerocercoids (Healy and Juranek, 1979). The organism is indigenous throughout many parts of the USSR; the Baltic Sea countries; central and southeastern Europe; Lake N'gami, Africa; northern Manchuria and Japan; and New South Wales, Australia. In the Americas it is found in northern Minnesota, extensive areas of Canada and Alaska, and in the lakes of southern Chile and Argentina.

Control of disease caused by cestodes is dependent on thorough cooking of fish from endemic areas; cysts are also killed by freezing at -10°C for 24 hours (Healy and Juranek, 1979).

Nematodes

Disease from *Anisakis simplex* arises in humans when improperly cooked, smoked, or raw fish are eaten (Cheng, 1976). Anisakids are usually found in herring less than 12 centimeters (cm) long. Although the medical literature contains few reports of anisakiasis in the United States, over 3,000 cases are reported annually from Japan (Fontaine, 1985).

Whitish to clear larval worms approximately 1 cm long are found in the posterior body cavity of fish, coiled and encapsulated (McGladdery, 1986; Smith and Wooten, 1975). About 2-3% of the total bore into muscle. Some of these may be loosely encapsulated. Worms are usually seen on fresh fillets or about anal areas in the round of the fish. Up to 30% of European herring may be infected in a single sample, but estimates of total numbers with actual fillet infection are as low as 0.001%.

It has been suggested that the prevalence of muscle infection is related to the intensity of body cavity infection. Fish with eight or fewer organisms in the body cavity seldom show fillet infection.

From 1958 to 1980, 13 cases were reported in the United States, for a mean of 0.59 case/year; however, 37 cases have been reported since 1980 (4.1 cases/year) (Deardorff et al., 1986; McKerrow et al., 1988).

Pseudoterranova dicipiens and *Contraceacum* closely resemble *A. simplex* . Both have been identified in pinnipeds in the Atlantic and northern Pacific. Cases in the United States associated with *P. dicipiens* have been reported on the West Coast, resulting from consumption of salmon and Pacific rockfish (*Sebastes*) (Kliks, 1983; Margolis, 1977; Myers, 1979).

The risk of infection from nematodes is reduced when fish are gutted soon after capture; salting may also decrease incidence. However, neither smoking nor light salting of fish normally affects larvae. Freezing at -20°C for 72 hours or heating above 55°C for 10 seconds kills adult parasites (Khalil, 1969). Because infection of fillets may be related to body cavity burdens, subsamples that cannot be immediately frozen or gutted could be checked for numbers of parasites. If the number exceeds 15, the fish should be banned for human consumption (Bier, 1976; Bier et al., 1987).

Other Nematodes

Angiostrongylus cantonensis is acquired by eating raw snails (the intermediate host for the organism). Crabs, prawns, and fish that have eaten infective snails can serve as transport hosts. Heavy infections cause an eosinophilic meningoencephalitis. Epidemics and sporadic infections occur most commonly in the South Pacific, Southeast Asia, and Taiwan (Alicata, 1988; Kliks et al., 1982). In recent years the parasite has spread to the Caribbean, and in 1986 it was reported in rat and snail populations in Puerto Rico (Anderson et al., 1986).

Gnathostoma spinigerum is acquired by eating raw or undercooked freshwater fish. The organism presents a clinical picture of larval migrans, a granulomatous lesion, or a stationary abscess. Infections have been reported from eastern and Southeast Asia, India, and Israel (Healy and Juranek, 1979).

PATHOGENS ASSOCIATED WITH FECAL POLLUTION OF THE MARINE OR FRESHWATER ENVIRONMENT

Species of fish and shellfish harvested from inshore waters that are subject to contamination by human or terrestrial animal feces, and by other industrial/agricultural pollutants, may contain bacteria and viruses that are pathogenic for humans (Liston, 1980). This is especially true for filter feeders that concentrate bacteria and viruses present in polluted waters, and oysters harvested from growing waters contaminated with human sewage have been associated with many outbreaks of enteric disease (Son and Fleet, 1980).

Viral Human Enteric Pathogens

More than 100 enteric viruses can be found in human feces. Families of viral pathogens associated with pollution of harvesting waters include picornaviruses, reoviruses, adenoviruses, caliciviruses, astroviruses, and unclassified viruses such as Norwalk and Norwalk-like viruses, Snow Mountain agent, small round viruses, and non-A, non-B hepatitis virus (NANB). Of these enteric viruses, only hepatitis A virus (HAV), caliciviruses, astroviruses, Norwalk virus, Snow Mountain agent, and NANB enteral hepatitis virus have been documented to cause seafood-associated illness (Bryan, 1986; CDC, 1989; Cliver, 1988; Gerba, 1988; Richards, 1985, 1987; Rippey and Verber, 1988).

These human enteric viruses are species specific and even receptor specific for certain cells. Once released into the marine environment they do not multiply, and their survival and persistence are based on many factors, including temperature, salinity, ultraviolet inactivation from sunlight, and the presence of organic solids or sediments. Of these factors, the most important are temperature lower than 10°C and the protective action of organic materials (Gerba, 1988). Human enteric viruses have been isolated in field studies only from molluscan shellfish and blue crabs taken from a sludge dump in the North Atlantic. Other marine animals, including lobsters, sandworms, detrital feeding fish, and conch, have been shown to take up enteric viruses when marine waters were experimentally seeded with these viruses, but viral uptake from naturally polluted water has not been reported in field studies. The only seafood implicated to date in the transmission of enteric virus from contaminated estuarine waters has been molluscan shellfish (Gerba, 1988).

Hepatitis Type A (Enterovirus Type 72)

Hepatitis A virus is a member of the family Picornaviridae. Of the nearly 200 human-adapted picornaviruses, 69 inhabit the enteric tract (Gerba, 1988; White and Fenner, 1986). Enteroviruses of the family Picornaviridae include the following species:

Polioviruses	(Types 1, 2, and 3)
Echoviruses	(Types 1-34; no 10 or 28)
Coxsackieviruses	(Types A1-A24 and B1-B6; no A23) (67 types)
Enteroviruses	(Types 68-71)
Enterovirus type 72	(Hepatitis A, HAV)

Of these enteroviruses, only HAV has been documented as a cause of seafood-associated illness, and has been isolated from infected seafoods and waters contaminated with human feces. For this reason, the other enteroviruses are not discussed here.

All of the enteroviruses are resistant to an acidic pH, proteolytic enzymes, and bile salts in the gut. Hepatitis A is less acid stable than other enteroviruses, but more heat stable, surviving at 60°C for 4 hours (White and Fenner, 1986). HAV is more chlorine resistant than indicator bacteria and the other enteroviruses, with the exception of Norwalk virus (Grabow et al., 1983; Keswick et al., 1985; Peterson et al., 1983). Hepatitis A is one of the most serious seafood-associated viral infections, causing a protracted and sometimes severe disease, but its mortality is only approximately 0.6%.

Hepatitis A virus is spread by the fecal-oral route. It is hyperendemic in countries that are overcrowded, and have inadequate sanitation and poor hygiene. Most infections in these communities occur in childhood and are subclinical. In more developed countries the disease is seen most often between the ages of 15 and 30 (Cliver, 1988; White and Fenner, 1986).

Contaminated food or water and person-to-person contact are the main routes of transmission of HAV. Each year, 20,000 to 30,000 cases are reported to the CDC. Of these cases, approximately 140 are known to be due to foods (0.5% of the total). Most of these food-borne outbreaks are due to inappropriate personal hygiene of infected individuals (Cliver, 1988). Outbreaks can also occur due to inadequate cooking of contaminated foods and human fecal contamination of drinking water supplies, swimming waters, and shellfish growing waters.

In the 1950s the first documented case of shellfish-associated HAV occurred in Sweden (Cliver, 1988). The first shellfish-associated case in the United States was documented in the 1960s (Cliver, 1988;, Gerba, 1988; Richards, 1985). Richards (1985) reported approximately 1,400 cases of molluscan shellfish-associated HAV from 1961 to 1984. Between 1978 and 1987, the CDC food-borne surveillance system (CDC, 1989) reported nine outbreaks involving 125 cases of seafood-related HAV; however, two of the outbreaks involving 92 cases were due to contamination from food handlers; NETSU reported approximately 45 cases of shellfish-associated HAV in the same period (Rippey and Verber, 1988). Richards (1985) noted that the incidence of molluscan shellfish-associated HAV has decreased in the last decade, although it is difficult to describe the trend of reported HAV infections in the United States with a single, unidirectional line. There is currently no valid indicator of human enteric viruses such as HAV in shellfish growing waters.

Caliciviruses and Astroviruses

Caliciviruses, like all enteric viruses, are found in human feces and are responsible for many cases of gastroenteritis. Transmission routes of calciviruses are the same as those discussed for other enteric viruses. Infection by calicivirus tends to cause a mild gastroenteric illness. No seafood-associated outbreaks due to these enteric viruses have been reported to CDC or NETSU (CDC, 1989; Rippey and Verber, 1988).

Unclassified Viruses

Unclassified viruses include poorly characterized agents of gastroenteritis such as Norwalk and Norwalk-like agents, Snow Mountain agent, small round viruses (SRVs), and non-A, non-B enteral hepatitis (Gerba, 1988; White and Fenner, 1986). Norwalk and Norwalk-like viruses resemble caliciviruses and are considered possible members of this family of viruses by some scientists (White and Fenner, 1986).

Outbreaks of viral gastroenteritis due to the Norwalk agent have been associated with swimming in waters contaminated with human sewage, eating food or drinking water that is fecally contaminated, and consuming uncooked or partially cooked shellfish harvested from estuaries contaminated with human fecal material. The first documented shellfish-associated outbreak of gastroenteritis involving Norwalk virus occurred in 1979 in Australia, where more than 2,000 people were involved (Gerba, 1988). Since then, many outbreaks of Norwalk or Norwalk-like viral gastroenteritis have been reported in the United States (CDC, 1989; Morse et al., 1986; Richards, 1985, 1987; Rippey and Verber, 1988). Norwalk and Norwalk-like viral illnesses associated with shellfish are a continuing problem; reported incidents have increased during the last decade (Richards, 1985). Between 1978 and 1987, NETSU reported 11 shellfish-related cases of Norwalk gastroenteritis and 71 cases of Snow Mountain agent (Rippey and Verber, 1988). CDC (1989) reported two outbreaks involving 42 shellfish-associated cases of Norwalk and related viruses. Richards (1985) reported an outbreak of shellfish-associated gastroenteritis involving 472 cases in Louisiana. Norwalk was suspected but not serologically documented. Richards (1985) also reported more than 6,000 shellfish-associated cases of unspecified gastroenteritis during the past 50 years. It is probable that many of these are of viral etiology, possibly Norwalk or Norwalk-related agents. More than 75% of these cases have been reported since 1980, which may indicate increased awareness and reporting of shellfish-related illnesses or an actual increase in infection rates. Norwalk virus is more resistant to chlorine than bacterial indicators and other enteroviruses (poliovirus and reovirus), and there is no indicator for human enteric viruses in shellfish or their growing waters (Grabow et al., 1983; Keswick et al., 1985; Peterson et al., 1983).

The etiologic agent of enteral or epidemic NANB hepatitis is unknown and may represent a group of related viruses rather than a single agent. It has been suggested that the NANB agent could be an enterovirus – possibly a different serotype of HAV (Overby et al., 1983). Enteral NANB hepatitis can be more severe than HAV, and infection by the NANB agent is associated with a high incidence of cholestasis. However, only one case of shellfish-associated NANB was reported to NETSU from 1978 to 1987; CDC did not report any seafood-associated NANB in the same period.

Enteral NANB hepatitis is transmitted mainly by sewage-contaminated water and sporadically by person-to-person contact. In the Middle East and Africa, it appears to be endemic (Overby et al., 1983). Cliver (1988) reported water-associated outbreaks in India, Africa, the USSR, and Mexico. Gerba (1988) reported enteral NANB hepatitis cases associated with consumption of raw shellfish in the United States. One incident of shellfish-associated NANB and 1,645 cases of shellfish-associated nonspecified (i.e., unknown etiology) hepatitis were reported by NETSU between 1978 and 1987 (Rippey and Verber, 1988). No cases of NANB or unspecified hepatitis reported to CDC between 1978 and 1987 were associated with seafood consumption. If an unspecified type of hepatitis is accurately reported as associated with seafood consumption, it would most likely be hepatitis A or enteral NANB. However, accurate case histories for an infection with such a long incubation (4 to 6 weeks) are usually difficult to obtain.

Disease Control for Human Enteric Viruses

Proper classification of shellfish growing waters based on valid human enteric virus indicators, as well as implementation and maintenance of proper treatment and disposal of sewage to prevent human enteric virus contamination of shellfish harvesting waters, are the most effective measures to deter raw shellfish-associated infections by these viruses. New sewage treatment methods may have to be developed. Norwalk virus and HAV are more resistant to chlorine than bacterial indicators (*Escherichia coli, Streptococcus faecalis*, and acid-fast bacteria) and other enteroviruses (poliovirus and reovirus), and there is no indicator for human viruses in shellfish or in growing waters. The only potential viral indicator that shows similar chlorine resistance is f2 bacteriophage. The specifications for disinfection of drinking water – free chlorine residue of 1-2 milligrams per liter (mg/L) for 1-2 hours at pH less than 8 and turbidity less than 1 unit – may not be sufficient to totally inactivate all HAV and is not sufficient for Norwalk virus (Grabow et al., 1983; Keswick, 1985; Peterson et al., 1983).

Effective enforcement to eliminate recreational and illegal ("bootlegging") harvesting and sale of raw molluscan shellfish from known sewage-contaminated growing areas must be developed and adequately funded.

Hepatitis A virus could also be controlled by vaccination, because infection results in permanent immunity. Although the U.S. Army is conducting live vaccine trials for HAV at this time, no licensed vaccine is available to the general public. Passive immunization with gamma globulin following known exposure to HAV is currently the primary method of control and prevention (Cliver, 1988; White and Fenner, 1986).

PROCESSING, DISTRIBUTION, AND PREPARATION-RELATED PUBLIC HEALTH HAZARDS

In addition to bacterial, helminthic, and viral pathogens that may be present in or on seafoods because of their presence in harvest waters, certain bacterial and viral pathogens can also contaminate seafoods during processing, distribution, and preparation. Some microorganisms pose risk to consumers of seafood from both harvest and postharvest sources, for example, hepatitis A virus and *Shigella*.

Bacterial Pathogens

Salmonella

Outbreaks of seafood-borne salmonellosis were reported to CDC between 1978 and 1987; however, no cases of salmonellosis associated with shellfish consumption were reported to NETSU in the same period. Typhoid fever, due to infection by *S. typhi*, is of only historic importance as a seafood-related hazard in the United States, but continues to occur in foreign countries, potentially impacting the safety of imported seafood products.

Salmonellae (nontyphoidal) have been isolated from a variety of fish and shellfish. The FDA Pathogen Surveillance Program documented sporadic *Salmonella* contamination of domestic cooked shrimp and imported raw lobster, raw shrimp, and miscellaneous fishery products in fiscal year 1988. In a study in Florida, approximately 20% of oysters, clams, and crabs were contaminated with *Salmonella*, with positives found during all seasons. However, no salmonellae were recovered from samples of mullet, the only free-swimming species studied (Fraiser and Koburger, 1984). The public health impact of these findings remains to be determined.

TABLE: Seafood-Associated* Salmonella *Outbreaks, 1978-1987

Year	State	Seafood	Location	Serotype
1978				
1979				
1980				
1981	Connecticut	Shellfish	Cafeteria	*typhimurium*
	Pennsylvania	Shellfish	Restaurant	*thompson*
1982	New Jersey	Seafood combination	Restaurant	*enteritidis*
	New York	Tuna macaroni salad	Restaurant	*enteritidis*
1983	New Jersey	Shrimp cocktail	Restaurant	Untyped
1984				
1985	Pennsylvania	Stuffed sole	Church	*enteritidis*
1986	Pennsylvania	Gefilte fish	Home	*braenderup*
1987				*sandiego*

SOURCE: CDC (1989).

Salmonella contamination of foods of animal origin is not uncommon, and human infections associated with a wide range of food vehicles are frequently reported. Estimates of infectious doses for *Salmonella* range from less than 100 to more than 1,000,000 organisms; characteristics of the host, the organism, and the vehicle are all important mitigating factors in the number of *Salmonella* needed to infect and cause disease.

Epidemiology and Risk Assessment

For 1973-1986, an average of 55 food-borne outbreaks of nontyphoidal *Salmonella* infections, affecting a total of 3,944 persons, were reported each year to the CDC Foodborne Disease Outbreak Surveillance Program. During the same 14 years, an annual average total of 32,957 and 35,490 total *Salmonella* cases were reported through the laboratory-based *Salmonella* surveillance system and the *Morbidity and Mortality Weekly Report* (MMWR), respectively.

From 1978 to 1987, CDC reported six seafood-borne outbreaks involving 147 cases. Three of these outbreaks involving 80 cases were shellfish associated. Two shellfish-associated outbreaks of confirmed nontyphoidal salmonellosis were reported by NETSU between 1894 and 1988: a 100-case outbreak in Florida in 1947 due to contaminated oysters, and a 22-case outbreak in New York in 1967 associated with oysters imported from England. No cases of shellfish-associated *Salmonella* infections were reported to NETSU from 1978 to 1987 (Rippey and Verber, 1988). Two outbreaks of nontyphoidal salmonellosis associated with salmon prepared by a caterer have been reported (Cartwright and Evans, 1988).

The salmon was epidemiologically and microbiologically implicated, and multiple food-handling errors were identified. The origin of contamination of the fish was not determined, although raw chicken prepared at the same time as the salmon was also positive for *S. montevideo*, the outbreak serotype. Consumption of raw shellfish harvested from sewage-polluted waters also has resulted in *Salmonella* infections (Flowers, 1988a). Although CDC and NETSU food-borne surveillance data and literature reports indicate that seafood is a much less common vehicle for *Salmonella* than are other foods such as chicken or red meat, fish and shellfish may be responsible for at least a small proportion of the total number of *Salmonella* cases that occur each year in the United States. However, current data are inadequate to make any attempt at estimating attributable risk.

A number of years have passed since raw molluscs were associated with an outbreak of typhoid fever in the United States (Bryan, 1980; Rippey and Verber, 1988). One seafood-associated outbreak involving 25 cases was reported to CDC in 1973, but shellfish were not implicated as the vehicle. The last shellfish-associated typhoid outbreak reported in the NETSU data base occurred in 1954 (Rippey and Verber, 1988). The risk of typhoid fever associated with imported seafood may vary, depending on the rates of infection in the country of origin and the degree of pollution of harvest waters. Recent sporadic cases have been reported in consumers of raw shellfish in the Mediterranean area (Caredda et al., 1986; Torne et al., 1988) and, based on the incidence of typhoid fever in local populations, the risk associated with products imported from emerging nations would appear to be elevated (Parker, 1984).

Disease Control

The microbial quality of harvest waters does not appear to be a good predictor of nontyphoidal *Salmonella* contamination, because oysters removed from closed and open beds had the same level of contamination (4%), and no correlation was observed between the presence of *E. coli* or fecal coliforms and *Salmonella* in finfish or shellfish (D'Aoust et al., 1980; Sobsey et al., 1980). Salmonellae do not appear to grow well in oysters during storage and are not selectively retained during depuration or relaying, so contaminated oysters can be purified (Son and Fleet, 1980). As with other pathogens, the risk of infection can also be minimized by ensuring that shellfish are cooked and handled properly.

Campylobacter Jejuni

One outbreak of shellfish-associated campylobacteriosis has been reported to NETSU. *Campylobacter* has been isolated from fresh and coastal waters, and Sydney rock oysters and raw clams have been implicated as the source of human campylobacteriosis (Arumugaswamy and Proudford, 1987). The infectious dose for *C. jejuni* may be as low as 500-800 organisms, according to studies of volunteers drinking contaminated milk (Black et al., 1983).

The predominant symptoms of campylobacteriosis are diarrhea, abdominal pain, fever, nausea, and vomiting (Blaser and Reller, 1981). At the mild end of the spectrum, symptoms may last for only 24 hours; conversely, *C. jejuni* may also cause relapsing colitis that mimics ulcerative colitis or Crohn's disease. Most patients have a relatively mild illness and recover without sequelae.

Since the first outbreak of food-borne campylobacteriosis was reported to the CDC in 1978, state health departments have reported an additional 50 outbreaks, affecting a total of 1,717 persons. Nationwide laboratory-based *Campylobacter* surveillance was initiated in 1982. By 1986, 39 states were enrolled and 10,066 cases reported. One domestic shellfish-associated outbreak of campylobacteriosis was reported by NETSU between 1978 and 1987, an outbreak due to contaminated hard clams that affected 16 persons in New Jersey in 1980.

In addition, *Campylobacter* was suspected in several outbreaks reported to NETSU in which the etiological agent was listed as unknown (Rippey and Verber, 1988). Because outbreaks of campylobacteriosis have been associated with drinking contaminated surface water, freshwater species of fish and crustaceans may also pose a risk. Thus, although the risk of *Campylobacter* infection associated with eating raw shellfish appears low, the data are inadequate to make any assessment of attributable risk.

Control of *C. jejuni* is facilitated by the organism's lack of resistance to environmental factors (Franco, 1988). It is sensitive to drying, oxygen, low pH, and heat. Its sensitivity to sodium chloride limits the importance of marine species as vehicles of campylobacteriosis (Doyle and Roman, 1981, 1982). However, packaging in modified atmospheres and long-term storage at poorly controlled refrigeration temperatures may increase the risk of seafood-borne campylobacteriosis.

Escherichia Coli

There are currently no data to indicate that seafood is an important source of *E. coli* infections in this country. Most infections by *E. coli* appear to be related to contamination of food or water, with some person-to-person amplification under unhygienic conditions (Gangarosa, 1978; Gross, 1983). Pathogenic strains (with the possible exception of enterohemorrhagic *E. coli*) are a much more important cause of diarrheal disease in the developing than in the developed world; thus, imported products may present greater risk than domestic seafood.

Data are inadequate to formulate any risk management plan for diarrheagenic *E. coli* – or to say whether such a plan is needed. Because *E. coli* is not selectively retained as gut flora in oysters, depuration or relaying may be helpful in reducing the risk of raw shellfish-associated human infections (Son and Fleet, 1980).

Yersinia Enterocolitica

No outbreaks of seafood-related yersiniosis have been reported to CDC or NETSU; however, *Y. entercolitica* has been isolated from finfish, mussels, oysters, and other foods (Lee, 1977; Morris and Feeley, 1976). In a study of finfish and shellfish from Puget Sound, 3.8% of fresh seafood sampled in retail markets was positive for *Y. enterocolitica* (Abeyta, 1983). Bacteria have also been isolated from oysters (13%), shrimp (4%), and blue crabs (21%) harvested from coastal waters off Texas (Peixotto et al., 1979).

Y. enterocolitica is found worldwide, not only in feces but also in food, water, and the environment (Lewis and Chattopadhyay, 1986; Morris and Feeley, 1976). It is commonly found in specimens from swine slaughterhouses, and the consumption of raw sausage made from tonsil-containing pork scraps has been associated with disease (Tauxe et al., 1987). The significance of *Y. enterocolitica* as a fish-or shellfish-associated pathogen remains to be determined. Because it can grow at refrigeration temperature (i.e., is a psychrotroph), pathogenic strains that contaminate seafood could reach an infectious dose during extended storage. No specific surveillance system is in place at CDC for yersiniosis. From 1978 to 1987, state health departments reported four food-borne outbreaks of yersiniosis to CDC; however, none was associated with seafood. Likewise, NETSU reported no shellfish-associated outbreaks of yersiniosis between 1984 and 1988 (Rippey and Verber, 1988).

There are inadequate data to formulate a risk management plan for *Yersinia*. Because most environmental strains appear to be nonpathogenic, the ability to identify potential pathogens by testing isolates for pyrazinamidase activity and plasmid-associated traits such as calcium dependence, autoagglutination, or Congo Red binding will be needed in any control programs (Stern, 1982). Recent studies suggest that pathogenic strains can also be identified by using one of several DNA probes (Cornelis et al., 1987; Miliotis et al., 1989; Robins-Browne et al., 1989).

Listeria

Listeria monocytogenes is very widespread in nature and has been observed in the feces of man, farm animals, and wild birds. For example, up to 26% of sea gulls are carriers of the organism. It has also been isolated from pond-reared trout and crustaceans. *Listeria* is also widespread in the environment and can be found in soil, vegetation, crop debris, and of course, fecal material. It appears to be part of the saprophytic microflora of grass and other plants.

Listeria monocytogenes causes a wide range of diseases in humans, the most important of which are meningitis, septicemia, and perinatal disease (Lamont et al., 1988; Newman et al., 1979). Other manifestations include brain abscess, cranial nerve palsy, diffuse encephalitis, endocarditis, pneumonia, hepatitis, and arthritis. Aside from pregnancy, other predisposing conditions include immunosuppressive disorders and extremes of age, although invasive listeriosis occasionally occurs in otherwise healthy, nonpregnant persons.

Epidemiology and Risk Assessment

Contaminated food is becoming increasingly recognized as an important vehicle of *L. monocytogenes* (Ciesielski et al.,

1987; McLauchlin, 1987). In recent years, *Listeria* has been implicated in outbreaks involving contaminated cabbage in Nova Scotia, fresh cheese in California, and pasteurized milk in Boston (Fleming et al., 1985; Linnan et al., 1988; Schlech et al., 1988). An outbreak of perinatal listeriosis was tentatively attributed to consumption of raw finfish and shellfish in New Zealand, although it is not clear whether patients consumed raw seafood more often than persons who did not become ill (Lennon et al., 1984).

Catfish fillets are often contaminated with *L. monocytogenes* from their environment, but no human cases of listeriosis have been attributed to the consumption of catfish. It has been difficult to estimate the extent of listeriosis in the United States; prior to 1986, listeriosis was not reportable. A recent study identified 229 cases from the 34 million persons living in the study areas, providing minimum estimates of 1,600 culture-confirmed cases of listeriosis and 400 deaths per year for the United States (Schwartz et al., 1988). A case control study performed in conjunction with the active surveillance project identified consumption of raw hot dogs and undercooked chicken as risk factors for listeriosis; questions on seafood consumption were not included on the questionnaire (Schwartz et al., 1988).

An active listeriosis surveillance project being conducted by CDC and selected state and local health departments includes questions on seafood consumption by cases and controls, and may help define the relative importance of seafood in the epidemiology of human listeriosis. The FDA found *L. monocytogenes* in seven of eight categories of seafood cultured in fiscal year 1988 (FDA, 1988). Therefore, *L. monocytogenes* may be an unrecognized cause of food-borne enteric disease.

Disease Control

L. monocytogenes is not easily inactivated by environmental influences, such as sunlight or freezing and thawing. It is found in soil, vegetation, sewage, and streams. Its prolonged presence in soil and on vegetation does not appear to depend on continual reintroduction but, rather, results from *Listeria*'s ability to exist as free-living forms on plants and in soil.

Temperatures for the growth of *L. monocytogenes* range from 1 to 45°C, with the optimum between 30 and 37°C. At 4-5°C its growth rate is 50 times slower than at 35-37°C, but the organism can still grow. The optimum pH for growth is near neutral to slightly alkaline pH, with a range of 5.0-9.6. *L. monocytogenes* is quite tolerant to salt and low water activity. It can grow in the presence of 10% sodium chloride and can survive at 25% sodium chloride (Bryan, 1979; Medallion Laboratories, 1987; Smith et al., 1987).

L. monocytogenes is fairly heat resistant, having $D_{150°F}$ values ranging from 12.8 seconds in skim milk to 17.6 seconds in cream (Bradshaw et al., 1985, 1987). Internalization in phagocytes did not significantly increase heat resistance (Bunning et al., 1986). Currently, FDA requires that *L. monocytogenes* not be present in ready-to-eat seafood products such as crabmeat or smoked fish. The restriction does not apply to raw products that will be cooked before eating.

Clostridium Botulinum

Human botulism is relatively rare in the United States. However, the control and prevention of botulism are some of the most important considerations in food processing. *C. botulinum* is widely distributed in soil and in the aquatic environment (Dolman, 1964; NFPA/CMI, 1984; Sakaguchi, 1979). Botulism is generally caused by consumption of food in which *C. botulinum* has grown and produced its toxin. The onset of symptoms of botulism usually occurs within 12-36 hours after ingestion of the food, with a range of 2 hours to 14 days. In general, shorter onset times result in more severe symptoms. Early signs may include gastroenteritis, weakness, lassitude, dizziness, and vertigo.

These are followed by eye problems such as blurred vision, diplopia (double vision), dilated and fixed pupils, and impaired light reflex reaction (Sakaguchi, 1979). Fever is absent and mental processes are normal. The major cause of death is respiratory failure and airway obstruction. Because of the severity of the disease, most cases are likely to be reported, although milder cases may go undiagnosed.

Botulism is among the most serious food-borne illnesses, with a mortality among untreated patients up to 70%. Competent medical treatment can reduce this mortality rate to approximately 15%. Fortunately, botulism is a rare disease, but cases due to consumption of seafoods were reported to CDC every year from 1978 to 1987. Nearly all of the cases in this period were caused by various ethnic foods prepared under conditions of poor general hygiene by individuals who did not understand the dangers of the procedure followed.

Epidemiology and Risk Assessment

C. botulinum is classified into toxin types A through G based on the serological specificity of the toxin produced. Types A, B, E, and F have been reported as causal agents in human food-borne botulism resulting from the consumption of foods in which bacteria have grown and produced toxins (Hauschild, 1989). Although type E is known as the "fish botulism organism," types A and B have been implicated in botulism caused by seafood (Hobbs, 1976). Nonproteolytic strains of types E and B are readily isolated from some marine sediments. Because *C. botulinum* produces heat-resistant spores and requires anaerobic conditions for growth, botulism has been most commonly associated with improperly processed canned food (usually home canned). Among fishery products, semipreserved items including smoked, salted, and fermented fish have been identified frequently as botulism vehicles (Eklund, 1982; Lynt et al., 1982).

The CDC reported 38 cases of fish-associated botulism from 1978 to 1987. Of these cases, 23 were from ethnic foods prepared and consumed in Alaska, and only one involved improperly home-canned seafood. All cases occurred in the home or as a result of eating homemade seafood products.

No domestic case of botulism from commercially prepared seafoods has been reported in the United States in recent years, although canned salmon from Alaska has caused two incidents in Europe. This suggests that for most seafood consumers the present risk is negligibly low, but it is high for particular ethnic groups. During the 1960s, several outbreaks of botulism occurred from consumption of improperly processed smoked fish stored under conditions permitting the growth of *C. botulinum* (Pace and Krumbiegel, 1973). Improved processing methods have reduced this hazard, but safety still depends largely on low-temperature (<3°C) storage of the processed product.

Disease Control

TABLE: Seafood-Related Botulism Outbreaks and Cases, 1978-1987

Year	State	Seafood	Location	Cases
1978	Puerto Rico	Home-marinated fish	Restaurant	3
1979	Alaska	"Other" fish	Home	1
1980	Alaska	"Other" fish	Home	1
1981	California	"Other" fish	Home	2
	Idaho	"Other" fish	Home	1
	Hawaii	"Other" fish	Home	1
1982	Alaska	Fermented fish eggs	Home	2
	Washington	Home-canned salmon	Home	1
1983	Alaska	Salmon eggs	Home	1
1984	Alaska	Stink eggs[a]	Home	3
1985	Alaska	Stink eggs	Home	8
	Alaska	Stink eggs	Home	1
	Alaska	Fish eggs	Home	1
	New Jersey	Kapchunka[b]	Home	2
1986	Alaska	Fermented fish heads	Unknown	3
	Alaska	Fermented fish eggs	Unknown	1
1987	Alaska	?[c]	?[d]	1
	New Jersey	?	?	5
Total				38

[a] Alaskan fermented fish eggs produced by burial in the ground.
[b] Dry salted fish.
[c] Unsure of vehicle.
[d] Unsure of location.
SOURCE: CDC (1989).

Botulism may be prevented by inhibiting growth of the organism or by destroying it in the food. Canning processes are generally designed to destroy the heat-resistant spores of *C. botulinum*, and this process is very effective. Type E spores of *C. botulinum* are more heat sensitive than type A or most type B, and pasteurization processes used in seafoods are generally effective in destroying most of them. Thus, hot smoking at 82.2°C can destroy 10,000,000 spores of type E in 30 minutes (Cann and Taylor, 1979), and processes used for crabs usually eliminate type E strains. However, as a safety factor it is necessary to store such heat-treated products below 3°C to prevent germination and growth of surviving spores and production of toxin. Processes such as heavy salting or drying, which reduce water activity below 0.975 (i.e., >5% salt), and pickling to reduce pH below 5.3 are effective in preventing growth of type E *C. botulinum*.

Light salting and other mild forms of processing are ineffective in destroying *C. botulinum* spores and, if followed by vacuum packing or controlled atmosphere storage, may actually promote growth of the organism by destroying competitive bacteria. It is essential that such products be held below 3.3°C and cooked prior to eating (Emodi and Lechowich, 1969).

Shigella

Seafood-associated outbreaks and cases of shigellosis have been reported to CDC and NETSU. The normal habitat for shigellae is the intestinal tract of humans and nonhuman primates and, rarely, other animals. Therefore, like typhoid fever, shigellosis usually originates with a human excreter (Blake and Feldman, 1986). Human volunteer studies indicate that ingestion of fewer than 100 shigellae can cause disease (DuPont et al., 1969). An estimated 300,000 cases of shigellosis occur in the United States each year, not all of which are food borne. Case fatality rates have been as high as 20% among hospitalized patients.

For the years 1978-1987, an average of seven food-borne outbreaks affecting a total of 573 persons were reported each year to the food-borne disease surveillance system (CDC, 1989).

During this time, seven outbreaks involving 137 cases were seafood borne. Four of the seven outbreaks, involving 77 cases, were shellfish associated. During the same period, annual averages of 14,460 and 18,498 cases were reported through the laboratory-based *Shigella* surveillance system and the MMWR, respectively. In 1978-1987, NETSU reported 84 cases of shellfish-associated shigellosis, 11 in California and 26 in Arizona in 1979, and 47 in Texas in 1986.

Most outbreaks of food-borne shigellosis result from contamination of raw or previously cooked foods during preparation by an infected food handler with poor personal hygiene. In view of the ability of *Shigella* to survive in seafood, the risk of seafood as a vehicle of shigellosis should parallel that of other foods.

Although shigellae are readily killed by heat and low pH, under certain conditions they can survive for some time outside the host (Flowers, 1988b). Taylor and Nakamura (1964) reported that *S. sonnei* and *S. flexneri* could survive at 25°C in clams for more than 50 days and in oysters for more than 30 days. Because of the low infectious dose, consumption of shellfish harvested from contaminated waters can result in disease.

Staphylococcus Aureus

S. aureus is a common human-associated bacterium. It is one of the most frequent causes of food poisoning in the United States and is often introduced during food preparation. The human carrier rate is 6-60%, with an average of 25-30% of the population being positive for enterotoxin-producing strains. The illness is an intoxication that results from consuming food in which the bacteria has grown and produced toxin. The bacterial population must reach high levels (usually > 100,000 organisms per gram) for enough toxin to be accumulated and illness to result.

The main reservoir for *S. aureus* is human nasal passages, but it is also found on skin, hands, wounds, and cutaneous abscesses. Additionally, it can be isolated from air, dust, floors, and other environmental surfaces, and it survives well in the environment. *S. aureus* is frequently isolated from seafoods, especially those that are handled during processing.

The CDC reported two seafood-associated outbreaks of *S. aureus* involving 12 cases between 1978 and 1987. One of these outbreaks, which involved nine cases, was shellfish associated. The other outbreak, involving three cases, was associated with other seafood products. The outbreaks were attributed to infected food handlers (CDC, 1989).

The conditions that affect the growth of *S. aureus* are often quite different from those for toxin production. In general, the relative ranges for toxin production are narrower than those for growth. For example, the temperature range for growth of *S. aureus* is 7-45°C, with an optimum of 35-37°C. Toxin production is maximum at the optimum, but very little toxin is produced at lower temperatures (>20°C). The pH range for growth of the organism is 4.5-9.3, but enterotoxin production is limited to pH 5.1-9. The optimum pH for both growth and toxin production is near neutral. The water activity (partial pressure of water in a food, divided by the partial pressure of pure water at the same temperature) minima for growth and toxin production differ, being 0.85 and 0.93, respectively (Banwart, 1989).

Good sanitary measures are necessary to prevent the contamination of seafood products with *S. aureus* because the enterotoxins are heat resistant and can withstand temperatures above boiling for long periods of time (Bergdoll, 1979). Because its presence is nearly unavoidable in products handled by humans, many states have suggested guidelines for *S. aureus*. One hundred organisms per gram is the most common upper limit in these guidelines. The symptoms of *S. aureus* food poisoning include nausea, vomiting, abdominal cramps, and diarrhea. The onset time for symptoms averages 2-4 hours but may range from 0.5 to 24 hours. The illness usually lasts only 1-2 days (Bergdoll, 1979; Jay, 1986).

Viral Pathogens

To prevent and control hepatitis A virus in processing, distribution, and preparation of foods, food handlers should be carefully educated and trained for good hygienic practices, particularly hand washing and sanitizing following defecation.

Hepatitis A virus (HAV) is a pathogen mainly transmitted by unsanitary food preparation practices. Each year 20,000 to 30,000 cases of hepatitis A are reported to the CDC. Of these cases, approximately 140 are directly attributable to contaminated foods (0.5% of the total). From 1978-1987 only two fish-associated HAV outbreaks involving a total of 92 cases were reported to CDC, 7 cases of HAV in 1980 due to tuna served in the home and 85 cases of HAV from tuna salad in a New York restaurant. There are no documented cases of transmission of human enteric viruses other than hepatitis A from seafood products contaminated at the processing, distribution, or food handling level (Bryan, 1986; CDC, 1981a-c, 1983a,b, 1984, 1985, 1989; Rippey and Verber, 1988). Viruses do not replicate in seafood products, so temperature abuse is not a factor. However, handling of products by a person infected with enteric viruses could result in transmission by the fecal-oral route if poor personal hygiene is allowed (Cliver, 1988; Matches and Abeyta, 1983).

IMPACT OF PROCESSING TECHNOLOGY

Heat Treatment

Heat treatment is frequently defined in terms of *D*- and *F*-values, with subscripts indicating temperature: *D* is the number of minutes at the indicated temperature necessary to reduce a microbial population by 90%, and *F* is the lethality of a process expressed as minutes at the indicated temperature. A third term Z is the temperature difference associated with a tenfold difference in killing rate. Whereas achievement of commercial sterility (as in canned fish) generally requires a 12 × D treatment, pasteurization may require lesser treatment (commonly 4-6*D*).

Nearly all canned seafoods available to the U.S. consumer are fully processed, commercially sterile products that are shelf stable in their original containers. Their production is controlled by FDA or state agencies working with FDA and following a Hazard Analysis Critical Control Point (HACCP) system that has proved very effective. Because fish and shellfish are low-acid foods, the major safety concern involves *Clostridium botulinum* whose spores are extremely heat resistant. Processors must deliver a treatment of 12*D* or greater, which is ensured by requiring the regulating agency's approval of process regimes and strict recordkeeping by the companies.

No cases of botulism from commercially canned seafoods have occurred in the United States in recent years. A single outbreak in Europe due to Alaska canned salmon was shown to result from improperly functioning equipment related to the use of collapsed three-piece cans. This problem has been resolved by improved maintenance and quality assurance programs and by widespread adoption of two-piece extruded cans (NFPA/CMI, 1984; Thompson, 1982).

For equal safety, imported fully processed canned seafood products must be made by processes that meet domestic FDA requirements, including process approval, adequate recordkeeping, and employment of certified retort/operator supervisors.

Semiconserved canned fishery products may contain living microorganisms because they receive milder heat treatment in pasteurization (Delmore and Crisley, 1979; Eklund, 1982; Lerke and Farber, 1971). These products, which are frequently imported, depend on acidification, use of salt, and storage at reduced temperature for their stability and are generally marketed as delicatessen items kept under refrigeration.

Pasteurization is a term that refers to a mild heating process, usually below 100°C. By definition, the term indicates that the product is not sterile and therefore may continue to harbor microorganisms. Consequently, pasteurized products must be continuously refrigerated so that surviving microorganisms will not multiply too rapidly, shortening the product's anticipated shelf life and introducing hazards. The initial microbial population greatly affects the efficiency of the process. The most important application of pasteurization in the seafood industry is the treatment of pickled crabmeat, which is packed, sealed in cans, and held under refrigeration after treatment. The National Blue Crab Industry Association has recommended a national standard that effectively provides for the destruction of potentially hazardous bacteria that could grow during refrigerator storage. The process provides the equivalent of $F_{85°C}$ in excess of 31 minutes (NBCIA, 1984).

Pasteurization has been proposed for other products including shrimp (Lerke and Farber, 1971), crawfish, and smoked fish (Eklund et al., 1988). Since the FDA dropped the GMP regulations on the processing of smoked fish, there has been increasing concern regarding the potential for a botulism outbreak associated with smoked fish. The pasteurization process described by Eklund et al. (1988) has the potential to minimize the concerns associated with this product. Pasteurization was more effective for smoked fillets and steaks than for dressed fish.

Sous vide products are minimally processed foods, including seafood, that are being introduced into the U.S. market. The products are portion controlled, vacuum packaged in plastic pouches or rigid containers that are highly impermeable to oxygen and moisture, and then cooked in either a water bath or a high-humidity oven. Cooking temperatures for *sous vide* are usually far lower than those associated with pasteurization. Cooking may involve temperatures close to the desired maximum internal temperature for the product, and thus require a long cooking time, or products may be cooked quickly at temperatures considerably above the desired internal maximum. In either case, the principles that apply to pasteurization also apply to this processing technology. Products should be cooked to a desired *F*-value and cooled quickly.

The safety of *sous vide* products has been questioned because they are processed only minimally and do not contain preservatives to prevent microbial growth. Furthermore, the cooking process does not eliminate nonproteolytic types of *C. botulinum*, and there is some question as to whether it may allow other organisms such as *Listeria* to survive. The shelf life and safety of these products depend solely on refrigeration; therefore, it is critical that psychrotrophic pathogens not survive. In the United States, these items are presented as refrigerated, ready-to-eat, or heat-and-serve products. They are produced under controlled conditions, and caution is exercised during distribution. As of late 1989, the products were being sold only to food service establishments and were not being sold retail. This will probably change in the near future as demand increases. The FDA has limited the use of *sous vide* products to approved food processing operations and currently does not allow production at retail establishments such as grocery stores. It is important that they be produced under an approved HACCP program and that the principles of pasteurization be understood and applied to their production.

Modified Atmospheres and Vacuum Packaging

Modified atmosphere packaging (MAP) is defined as a process in which air is removed and replaced with other gases, usually carbon dioxide (CO_2) or a mixture of CO_2 and other gases. The atmosphere in the package will change over time because of microbial respiration and permeability characteristics of the packaging material. Controlled atmosphere packaging (CAP) is like MAP, except that the atmosphere is monitored and kept constant. For vacuum packaging, air is removed without replacement. It is well established that modified atmosphere (MA) storage of seafood products will extend the shelf life considerably by inhibiting the normal psychrotrophic microflora and reducing the spoilage rate. There is concern about the potential growth and toxin production of nonproteolytic types of *C. botulinum* (Garcia et al., 1987).

It has been demonstrated that MA packaging or vacuum packaging coupled with temperature abuse can create conditions in which toxin production occurs before sensorially detected deterioration (Genigeorgis, 1985; Post et al., 1985). The National Marine Fisheries Service (NMFS) recommends (but does not require) that fisheries products not be MA or vacuum packaged if they are to be stored under refrigeration (Post et al., 1985). Where vacuum or MA packaging is used, NMFS has set guidelines (NOAA, 1990).

PUBLIC HEALTH RISK ASSESSMENT

Available epidemiologic data are inadequate to permit quantitative assessment of the risk of illness due to infectious agents acquired by eating seafood. However, it is possible to develop some qualitative data on risk.

Risk Assessment

Hazard Identification

A variety of seafood-associated bacterial, viral, and parasitic agents cause illness in humans. These agents can generally be divided into three groups: (1) those that are known to cause disease in healthy adults; (2) those that usually do not cause disease in healthy adults but can cause illness in special population groups (children, immunosuppressed patients); and (3) those that are of uncertain pathogenicity for humans. It should be kept in mind that these categories are not absolute and that agents could be moved from one category to another in response to new information and changing conditions. Pathogens such as *Salmonella, V. parahaemolyticus*, or Norwalk virus cause disease in normal, healthy hosts; however, the illness in normal hosts is generally mild and self-limited. In contrast, when persons who are immunocompromised or otherwise at high risk are infected, severe, life-threatening illness may occur (e.g., *V. vulnificus*).

Dose-Response Assessment

Viral agents such as Norwalk and some bacterial agents such as *Shigella* require a very low infectious dose to cause illness (ca. 10^2 CFU for *Shigella* and theoretically one infectious viral particle for Norwalk).

These agents are often transmitted by direct fecal-oral contamination and may pose a problem if they are allowed to contaminate food items after processing. Other bacteria generally require a higher infectious dose, which may be reached by time/temperature abuse of food products. Unfortunately, in the absence of volunteer studies the infectious doses for many of the viral and bacterial pathogens discussed are unknown; and even if such studies were carried out, it could not be proved that study conditions duplicate the wide range of human responses to natural exposure. However, it is probably accurate to say that the risk of infection increases with increased dose for most pathogens and that the dose depends to a large degree on the handling of the product after harvest, during processing, and in preparation.

Exposure Assessment

The risk of exposure is dependent on a number of variables, including the type of seafood, the location, the water quality at harvest, and the way the product is handled after harvest. Risks are markedly different for each pathogen and each product, making it difficult to generalize about overall risk of exposure to infectious agents or to establish uniform water quality or product safety standards.

Infections with *Vibrio* species and other naturally occurring marine bacteria are generally associated with eating shellfish. Their numbers are dependent on salinity, temperature, and a variety of other factors, with temperature (>20°C) probably being the most important variable (Baross and Liston, 1970).

Of the enteric viruses, only hepatitis A virus, Norwalk virus, Snow Mountain agent, caliciviruses, astroviruses, NANB enteral hepatitis, and unspecified hepatitis have been documented to cause seafood-associated illness. With the exception of HAV contamination of ready-to-eat seafoods by food handlers, all reported cases of seafood-associated viral infections have been from the consumption of raw or improperly cooked molluscan shellfish (Bryan, 1986; CDC, 1989; Cliver, 1988; Gerba, 1988; Richards, 1985, 1987; Rippey and Verber, 1988). Human enteric viruses from naturally occurring contamination have been isolated in field studies only from molluscan shellfish and from blue crabs collected at a sludge dump in the North Atlantic.

Other marine animals, including lobsters, sandworms, detrital feeding fish, and conch, have been shown to take up enteric viruses seeded into marine waters experimentally, but this has not been reported from field studies to occur naturally. These viruses are very species specific and even receptor specific for certain cells. Once released into the marine environment, their survival and persistence are based on many factors, including temperature, salinity, ultraviolet inactivation from sunlight, and the presence of organic solids or sediments. Of these factors, the most important appears to be a temperature below 10°C and the protective action of organic material (Gerba, 1988).

Risk Characterization

Raw or undercooked shellfish appear to carry a higher risk of disease than do other types of seafood, particularly because such shellfish constitute less than 0.1% of the seafood consumed in the United States but are responsible for a large proportion of the cases reported (CDC, 1989; Rippey and Verber, 1988).

Available surveillance data from both CDC and NETSU and clinicopathophysiologic studies suggest that food-borne diseases due to unknown etiologies, unspecified (i.e., unknown etiology) hepatitis, and certain *Vibrio* species (*V. parahaemolyticus, V. vulnificus*, non-O1 *V. cholerae*) represent the greatest risk for persons consuming raw molluscan shellfish.

The largest number of reported seafood-associated illnesses have unknown etiologies (CDC, 1989; Rippey and Verber, 1988). Although these are probably not all of microbiological etiology, many have onset periods and symptoms consistent with the clinical pathology of Norwalk and related viruses.

However, both the CDC and the NETSU data bases report relatively few cases of confirmed Norwalk or related viral infections associated with shellfish because only a few specialized laboratories are able to serologically diagnose infections with Norwalk, Snow Mountain, and related viruses. Despite the low number of documented cases of Norwalk and Norwalk-like agents in shellfish, recent evidence suggests that shellfish-associated infections with these agents occur more frequently than identified (Morse et al., 1986) and that they may ultimately be the most common shellfish-associated pathogens, especially in coldwater clams.

The risk of infection by bacterial pathogens appears to be lower than that due to viruses. Although data are limited, the incidence of infection with the more common *Vibrio* species in coastal states (where shellfish consumption is probably highest) is in the range of 0.5-1.0 case/100,000 population/year (Hoge et al., 1989). Although cases of *V. vulnificus* associated with shellfish are reported with approximately the same frequency as *V. parahaemolyticus* and non-O1 *V. cholerae*, the severity of the infection and the high mortality rate in populations at risk elevate the importance of the former as a seafood-associated pathogen. Shellfish-associated infections due to *Salmonella* and *Shigella* are approximately the same, based on CDC surveillance data. However, cases due to these latter agents are probably underdiagnosed because the vehicle for the vast majority of cases caused by these agents is never identified.

Parasitic infections are less common than bacterial and viral infections in the United States, with *Anisakis simplex* and cestodes having the greatest public health significance. In general, parasitic infections are concentrated in certain ethnic groups that favor the consumption of raw or partially cooked seafood from "high-risk" geographic areas.

RISK MANAGEMENT

Control of the risk of infectious seafood-borne disease is generally achieved by (1) excluding the agents from food, (2) controlling their growth in food, or (3) destroying them. All three methods depend to some extent on the ability to detect the occurrence of these microorganisms and to follow procedures that achieve one or more of the objectives listed.

Pathogens in Marine or Freshwater Environments

Risks associated with Vibrionaceae can be addressed at several levels. Because these organisms are a "natural" part of the bacterial flora of bays and estuaries, their presence cannot be predicted by standard indicators such as fecal coliform counts. However, their occurrence is dependent on environmental conditions (water temperature, salinity), with counts generally peaking in the late summer and early fall. The risk of infection from Vibrionaceae could be minimized by restricting summer harvesting of oysters.

More realistically, a system for monitoring shellfish and harvest waters for these organisms could be established, analogous to the current system for monitoring fecal coliforms. Rapid diagnostic techniques for identification of pathogenic vibrios are available or are being developed (Kaper et al., 1988), making such a system technically possible. Monitoring data would provide an index to times when the risk of exposure to *Vibrio* is greatest and could serve as an adjunct to other indicators for the management of harvest waters.

Efforts can also be directed toward reducing or eliminating organisms present in shellfish after harvest. The self-cleansing of some enteric viruses and marine vibrios may not proceed at the same rate as for enteric bacteria and indicators, which suggests that current techniques for depuration may be of little benefit in limiting contamination with Vibrionaceae. New technology-enhanced depuration must be developed for eliminating naturally occurring vibrios (Rodrick, 1990). Low-dose gamma irradiation of live shellstock and fresh or frozen seafood products has also been shown to be an effective means of eliminating *Vibrio* species (Giddings, 1984). Dose levels of 1.0 kilogray (kGy) (100,000 rads) to 1.5 kGy (150,000 rads) will reduce high levels (10^6) of seeded *Vibrio* spp. in live shellstock oysters to undetectable levels. The LD_{50} of live shellstock oysters is 2.25 kGy (Kilgen et al., 1988). Sensory evaluation shows no significant organoleptic differences in shellstock oysters treated with 1.5-kGy gamma irradiation (Kilgen et al., 1988). Gulf Coast summer oysters naturally infected with *V. vulnificus* levels of 10^4 organisms per gram of oyster meat also had no detectable *V. vulnificus* levels after gamma irradiation of 1.0 kGy (M. Kilgen and M. Cole, Nicholls State University, Thibodeaux, La., personal communication, 1990).

Although data on human infectious dosages are not available for all of the Vibrionaceae, it is likely that the risk of infection can also be reduced by limiting the multiplication of organisms during handling and processing. Recent studies demonstrate that there can be a striking increase in bacterial counts (including counts of *Vibrio* species) when shellfish are left unrefrigerated on the decks of boats during warm summer days (Cook and Ruple, 1989). Similarly, bacterial counts can increase if shellfish are not adequately refrigerated during transportation to plants and subsequent distribution. Some processors are currently using rapid cool-down techniques to minimize these risks; such techniques need to be further evaluated, and if shown to be effective, their implementation should be encouraged.

Finally, the consumer can play an important role in minimizing risks. Consumers should be educated about the need for proper refrigeration of fish and shellfish, and the dangers of cross-contamination. They should be advised about the risks of eating raw shellfish. Of particular concern are persons who are at high risk for septicemia due to *V. vulnificus* (persons who are cirrhotic, who have hemochromatosis, or who are in some way immunocompromised). Such persons must not eat raw or undercooked shellfish.

With the development of gene probes amplified by polymerase chain reaction and other rapid, new diagnostic techniques, shellfish can be screened for the presence of some potentially pathogenic *Vibrio* species (Kaper et al., 1988). These probes could theoretically be used as part of a national monitoring system for shellfish. However, further studies are required to determine the ways in which data of this type can be used to protect the public health.

The risk of parasitic infection depends largely on the species of fish and shellfish and the area from which they are harvested. In general, prevention of infection is dependent on limiting the harvesting of seafood from high-risk areas; informing sports fishers of regionally important risks associated with certain species and the availability of alternate control strategies (e.g., freezing fish before raw consumption); and educating consumers on the importance of proper cooking of fish and shellfish.

Some parasitic agents, such as the anisakids, are detectable by visual inspection techniques. However, these are probably the only infectious agents that would be detected by such a system, and the very low frequency of reported disease due to these agents does not justify using public resources specifically to identify parasites in seafood.

Pathogens Associated with Pollution

Control of pollution-associated agents generally translates into identifying sewage contamination of harvest areas. The only seafood products that are regulated by microbiological indicator standards for growing water quality are molluscan shellfish. The sanitary quality of shellfish is based on an allowable standard of 14 most probable number (MPN) fecal coliforms per 100 milliliters (mL) of growing water, with not more than 10% exceeding 43 MPN fecal coliforms/100 mL (FDA,

1989a). The National Shellfish Sanitation Program (NSSP) manual of operations also requires a sanitary survey of oyster growing estuaries prior to approval for shellfishing, relaying, or depuration (FDA, 1989a). The revised NSSP recommendations (FDA, 1989b) for monitoring interstate shellfish shipments state that:

> Proper growing area classification and strict adherence to good manufacturing practices (GMP's) are the principal considerations for assuring the safety of shellfish. The bacteriological market guideline of 230 MPN fecal coliforms/100 g [grams] oyster meat should no longer be used as the primary basis for embargoing or destroying shellfish shipments. Shellstock shipments should be monitored within 24 hours of entering the state, and should be considered acceptable if: shipments are properly identified (by tag, bill of lading or label) and cooled, shellstock to 50°F or less, and shucked product to 45°F or less, and if they are in compliance with all other NSSP conditions of shipment. Shellfish should be rejected if shipments are not properly identified, are inadequately cooled (shellstock shipments exceeding 60°F and shucked product shipments exceeding 50°F), or other conditions exist, e.g. decomposition and/or adulteration with poisonous and deleterious substances. Shellfish received under conditions between acceptance and rejection criteria could be examined bacteriologically, and the following guidelines applied for acceptance: fecal coliform density of not more than 230 MPN/100 g and aerobic plate count (APC) of not more than 500,000/g.

The fecal coliform standard in growing waters and the guidelines for meats were established to prevent sewage contamination of shellfish that may be consumed raw. Fecal coliforms are present in high numbers in untreated sewage and are considered "indicators" of the possible bacterial and viral human enteric pathogens that may also be found in feces. The allowable numbers of fecal coliforms in shellfish growing waters were based on the relationship between total coliform counts and the number of *Salmonella typhi* present during a typhoid fever epidemic in the 1920s (FDA, 1989a). A conference was called in 1925 to establish the basic concept of a national program for sanitary control of the molluscan shellfish industry. The first growing water standard was 70 MPN/100 mL total coliforms with no more than 10% of samples exceeding 230 MPN total coliforms/100 mL and freedom from direct contamination with fresh raw sewage.

This indicator standard was established through epidemiological investigations from 1914 to 1925 by the states and the Public Health Service. It was believed at this time that typhoid fever would not normally be attributed to shellfish harvested from water in which "not more than 50 percent of the 1 cc [cubic centimeter] portions of water examined were positive for coliforms" (FDA, 1989a).

This equated to 70 MPN total coliforms/100 mL, which is equivalent to the fecal material from one person diluted in 8 million cubic feet of coliform-free water. Later studies concluded that fecal coliforms were more accurate indicators of fecal contamination than total coliforms, and the allowable numbers were extrapolated to the current standard of 14 MPN fecal coliforms/100mL growing water, with no more than 10% of samples exceeding 43 MPN fecal coliforms/100 mL (FDA, 1989a). However, even the more specific fecal coliform indicator of contamination in water and seafood has been questioned (Cole et al., 1986; Elliot and Colwell, 1985; Kilgen et al., 1988; Matches and Abeyta, 1983).

TABLE: State and Federal Microbiological Standards and Guidelines

FDA	Seafood Products	Total Aerobic Plate Count	Coliform	Escherichia coli	Salmonella	Coagulase+ *Staphylococcus aureus*
Administrative guidelines	Clams, mussels oysters (fresh or frozen)	500,000/g	-	230/100 g	-	-
Compliance policy guidelines	Crabmeat (fresh or frozen)	No standard	-	3.6/g	-	-
	Langostinos (frozen, cooked)	100,000/g	20/g	3.6/g	-	3.6/g
Import alert	Seafood	-	-	-	0	-

SOURCE: Martin and Pitts (1989).

According to Garrett (1988), the United States has only three types of general regulatory microbiological indicator criteria for other seafoods:

Microbiological standards

- FDA's GMP requirement that thermally processed low-acid foods packaged in hermetically sealed containers be commercially sterile
- FDA's microbiological standards for "approved," "conditionally approved," "restricted," or "prohibited" molluscan shellfish growing waters, used in conjunction with a sanitary survey as described above

Microbiological guidelines

- Microbiological criteria for crabmeat and langostinos
- Defect action levels for raw breaded shrimp plants
- Bacteriological market guidelines for molluscan shellfish (revised to GMPs for proper refrigeration, handling, and tagging in 1989) (FDA, 1989b)

Microbiological Specifications

- Microbiological acceptance criteria contained in federal specifications for the procurement of fish and frozen shelled oysters

The National Fisheries Institute published a draft *Handbook of State and Federal Microbiological Standards and Guidelines* (Martin and Pitts, 1989). The basic requirements for an ideal indicator have been discussed at length (Elliot and Colwell, 1985). Some of the characteristics necessary for an indicator to accurately predict the presence of sewage-associated pathogens include:

1. specificity for the source of the pathogens (i.e., raw sewage),
2. presence in feces in sufficiently high concentrations to allow sensitivity of detection,
3. persistence in the aquatic environment at least as long as the pathogens,

4. absence of growth in the aquatic environment, and
5. availability of inexpensive methods for detection and quantification.

A number of indicator organisms have been proposed in the past for detection of fecal pollution in fresh, brackish, estuarine, and seawaters. However, no single indicator organism exists for determination of public health risk in waters or seafood (Elliot and Colwell, 1985). Some of the problems encountered with the current fecal coliform indicator standard in growing waters and the guideline for oyster meat include the following:

Non-Escherichia coli fecal coliforms and even non-sewage-related bacteria may predominate in the fecal coliform population analyzed by methods approved by the American Public Health Association (APHA, 1985a,b). These non-E. coli fecal coliforms can be found in shellfish, sediments, and the water column, especially at warm temperatures (Cole et al., 1986; FDA, 1989b; Kilgen et al., 1988; Paille et al., 1987). Characterization of the non-*E. coli* fecal coliform population in Louisiana oysters showed that in the warm months, *Klebsiella pneumoniae* isolates accounted for 86% of the non-*E. coli* fecal coliforms and often outnumbered *E. coli* by 1,000 to 1 (Paille et al., 1987). These *Klebsiella* oyster isolates were further characterized and compared with *K. pneumoniae* human clinical isolates by electron microscopy, guanine:cytosine ratios, and antibiotic resistance. The results suggested that *K. pneumoniae* isolates from oysters were of environmental, not sewage, origin. Their seasonal variation was typical of environmental bacteria, and they did not exhibit the multiple antibiotic resistance characteristic of the clinical strains. Similar studies by the FDA agreed that Gulf of Mexico Coast oysters harvested from approved growing waters in summer months may contain excessively high levels of non-*E. coli* fecal coliforms and not represent an excessive health hazard (FDA, 1989a). It was concluded that fecal coliforms may not be a reliable indicator of fecal contamination in Gulf Coast oysters, especially in summer months, and that *E. coli* would be a better indicator for an oyster meat guideline (FDA, 1989b; Kilgen et al., 1988; Paille et al., 1987). This resulted in the adoption of a one-year interim oyster meat guideline of 230 MPN *E. coli*/100 g instead of 230 MPN fecal coliforms/100 g by the Interstate Shellfish Sanitation Conference (ISSC) in 1983. This is still recommended in the NSSP operations manual for shellfish from approved growing waters with excessively high fecal coliform indicator levels, but it is not an officially adopted guideline (FDA, 1989b).

The fecal coliform indicator for waters does not indicate the presence of non-sewage-related naturally occurring aquatic bacterial pathogens, such as Vibrionaceae (Elliot and Colwell, 1985; Matches and Abeyta, 1983; Tamplin et al., 1982).

The fecal coliform indicator does not correlate with the presence of human enteric viruses, which are the pathogens most commonly associated with sewage contamination of waters and seafood (Cole et al., 1986; Elliot and Colwell, 1985; Gerba, 1988; Kilgen et al., 1988; Richards, 1985, 1987; Sobsey, 1980).

The human enteric pathogens of main concern from sewage contamination are now enterovirus type 72 (hepatitis type A) and Norwalk or Norwalk-like gastroenteritis viruses. However, current data also indicate that a constant and predictable relationship does not exist among fecal coliform indicators, *E. coli* and enteric viruses in estuarine waters and shellfish.

The public health risk associated with fecal material from animal sources versus human sources is also in question. The most productive shellfish growing estuaries are often those most subject to rainfall runoff from animal non-point sources. Extensive closures due to high fecal coliform indicator counts from non-point animal sources have been identified as one of the major concerns of state regulatory agencies and industry members from coastal areas; a great deal of research is required to assess human health risks from wild and domestic animal runoff (Elliot and Colwell, 1985; Kilgen, 1989). Terrestrial mammals carry bacterial species pathogenic to humans; however, these have generally not been associated with shellfish-borne illnesses. Rather, sewage-associated human illnesses appear most frequently to have a viral etiology, and viruses tend to be species specific. Therefore, an indicator of human enteric viruses in water and in seafoods is needed. Some indicators that have been proposed include poliovirus type 1, enterococci, *E. coli*, coliphages, fecal streptococci, *Clostridium perfringens, Pseudomonas aeruginosa, Bifidobacterium* species, *Rhodococcus* species, *Streptococcus bovis, Bacteroides* phages, and F+ phages (Elliot and Colwell, 1985; Kilgen, 1989; Richards, 1985). None of the suggested indicators appears to be an adequate indicator of human health risk from enteric virus pathogens in seafoods of water, and none of them would predict the presence of naturally occurring bacterial pathogens such as the Vibrionaceae. This is an area for vital research.

Depuration (controlled purification) and relaying have been used successfully to remove enteric bacterial pathogens and indicators from molluscan shellfish (Richards, 1988). However, the self-cleansing of some enteric viruses and marine vibrios may not proceed at the same rate as for enteric bacteria. Hepatitis A virus persists far longer in oysters and clams than *E. coli* or poliovirus (Sobsey, 1990). Improperly depurated shellfish have been responsible for outbreaks due to enteric viruses (Richards, 1988). Inasmuch as marine vibrios may also be present far longer than standard indicators and can multiply quickly at elevated temperatures, immunocompromised individuals should not assume that depurated shellfish are safe to consume. More research is needed both to develop indicators of adequate viral depuration and to develop enhanced depuration technology to remove naturally occurring vibrios (Rodrick, 1990).

Pathogens Associated with Processing and Distribution

Methods involved in processing can increase or decrease bacterial populations present in seafood. Processing methods are amenable to monitoring by an HACCP system. However, good data are not always available to justify currently used endpoints. Further research is needed to determine the appropriateness of current quality control criteria and to demonstrate that the use of these criteria results in a decrease in the incidence of human disease.

The major problem associated with distribution of seafood products is time/temperature abuse. Some pathogenic microorganisms can grow to dangerous numbers when exposed to prolonged temperature abuse. A bacterium that may double every 30 minutes at 25°C may require 1,200 minutes at 1°C. Therefore, it is very important that seafoods be cooled quickly to achieve maximum shelf life and to maintain safety. Only four bacterial pathogens (*Listeria monocytogenes, Yersinia enterocolitica, Aeromonas hydrophila,* and nonproteolytic *Clostridium botulinum* types E, B, and F) are considered psychrotrophs and are capable of growing at refrigeration temperatures (<4°C). Of the organisms described, only the nonproteolytic *C. botulinum* types have been definitely implicated in seafood-related illness. Time/temperature abuse during transportation was the major factor in outbreaks in the 1960s when smoked fish was transported unrefrigerated from the Great Lakes to distant markets (Pace and Krumbiegel, 1973).

Other bacterial pathogens that can grow at near-refrigeration temperatures (between 4 and 10°C) include enteropathogenic *E. coli, Staphylococcus aureus*, salmonellae, and *Vibrio parahaemolyticus*. These microorganisms may become a problem during transportation if seafoods are not cooled properly at any time after harvest.

Proper cooling of seafoods for transportation is critical for safety. The seafood must be cooled before being placed in the distribution system. Most refrigerated or frozen product transport systems are designed only to maintain a predetermined temperature and do not have the capacity to reduce the temperature of the product significantly. Cooling can be a lengthy process with large lots of fish, and temperatures can remain in the above 10°C range for many hours, permitting the growth of potentially hazardous bacteria if the warm product is loaded into transport containers. Chilled product should be loaded at internal temperatures below 4°C and frozen products at or below -18°C.

Imports

A large part of imported seafood originates in countries with a much higher incidence of enteric disease and lower levels of hygiene and sanitation than the United States. Therefore, seafoods from these countries could be contaminated more frequently or at a higher level with disease-producing organisms than our domestic fisheries product. It is important that the same principles of seafood-borne disease control be applied to these products as to domestically produced seafoods, preferably through an equivalent inspection system. This will require intelligent assessment of the microbiological hazards likely to be particularly troublesome in different exporting countries and the risks associated with in-country handling and processing practices. Because of the high risks associated with raw molluscan shellfish, the importation of shellfish for raw consumption should be prohibited unless standards for the microbial quality of harvest waters and postharvest processing in the exporting country are fully equivalent to those in the United States.

Future Risk

The fishing industry is in a period of rapid and profound change as a result of depletion of traditional stocks, development of domestic fisheries in Third World countries, exploitation of new fish resources, technological innovation, and changing consumer demands and expectations. The changes have had both beneficial and potentially adverse effects on microbiological risk to the final consumer of seafoods. Changes in canning technology, process recording equipment, temperature control, and freezing methods (leading to rapid cooling and freezing of much seafood at the point of capture) have undoubtedly improved microbiological safety for many seafood products, although this has not been definitively demonstrated. The public demand for lightly cooked and "fresh"-type products, which is satisfied by vacuum packaging and relatively long-term holding, can lead to potentially more hazardous products as discussed in relation to the *sous vide* process.

Control of temperature during the handling and storage of seafoods is widely recognized as critical for microbiological safety, but newer customer presentation systems require more stringent temperature control than those to which the industry has become accustomed. Safe storage of lightly cooked vacuum packaged products for more than a few hours requires temperatures consistently below 3°C and probably nearer to

0°C if they are to be stored for more than a few days (Post et al., 1985). These stringent conditions are necessary to prevent the growth of *Clostridium botulinum* type E, psychrotrophic pathogens such as *Listeria* and *Yersinia*, and possibly opportunistic pathogens such as *Aeromonas*.

The new seafood analogue products, which are widely used as a substitute for crab and other expensive species in seafood salads or other dishes, are manufactured from Alaskan pollock and a few other fish species. These products contain other food components such as sugars, emulsifiers, and egg white and are prepared by processes that remove most of the soluble "fishy" components from the raw material. They are heat treated so that most of the naturally present bacteria are destroyed. This provides a long shelf life under good storage conditions. These virtually sterile products provide an excellent growth medium for contaminant bacteria and do not develop the characteristically unpleasant odors associated with "bad" fish, which most consumers use as a warning not to eat the product. Therefore, care must be taken to avoid cross-contamination and warming of such products.

Ethnic Foods and Regional Considerations

The population of the United States includes many ethnic groups with different food preferences. In coastal regions this may be reflected in the patterns of fish and shellfish consumption. From a microbiological standpoint, the most important ethnic practices involve consumption of raw and fermented seafood products. Eating raw fish is most common among Americans of Asiatic origin and presents potential problems associated with parasites and *Vibrio* infection. As noted earlier, the increasing popularity of sushi and, to a lesser extent, seviche has spread the practice more widely among the U.S. population. So far, there has not been evidence of a major microbiological problem, but this situation must be watched carefully.

Fermented products present a more acute hazard because they are normally produced in the home by using traditional procedures that are poorly controlled. Most cases of seafood-borne botulism in the United States in recent years occurred in Alaska among native Americans as a result of consuming fermented fish and marine mammal parts. Little is known scientifically about the processes used, which seem to be largely proteolytic in nature; more study is clearly required in this area to identify procedures that would be safer than those presently used.

Traditional methods of producing marinated and cured fish may also present botulism problems, as evidenced by cases in Puerto Rico and New Jersey. In most such cases, proper temperature control during preparation could eliminate the problem. Education of specific targeted groups is necessary to manage these risks.

Other risks arise from the practice of consuming parts of fish, particularly intestines and other viscera, common among recent immigrants from Southeast Asia. In most cases, such dishes are eaten after cooking, which may explain the absence of major infectious disease outbreaks. However, the consumption of intestines and whole scallops has caused intoxications due to paralytic shellfish poisoning (PSP). It is unusual for scallop adductor muscle tissue to become toxic for humans, but viscera and other organs readily become toxic. Control has typically been based on measurement of toxin in the adductor muscle. Regulations and control measures designed to deal with common U.S. eating practices may not adequately protect individuals with different eating customs.

Aquaculture

The microbiological safety and promising future of the aquaculture industry has been reviewed by Ward (1989). Many of the microorganisms of concern in growing waters and abusive conditions in processing, handling, or preparation of other seafood products are also of concern in aquaculture. Two different areas of concern for aquaculture are pathogens known to cause disease in both fish and humans (e.g., *Edwardsiella tarda* and *Aeromonas hydrophila*) and the potential development of antibiotic resistance in these pathogens. The use of antimicrobials in aquaculture at both therapeutic and subtherapeutic levels is widespread (Brown, 1989). The possible development of antibiotic-resistant strains of bacteria, such as *Salmonella*, resulting from these feeding practices, and the subsequent transfer of such resistant bacteria on seafood products to humans are of concern.

Significant levels of resistance to tetracycline compounds, some of it transferable, have been documented in *A. hydrophila* strains isolated from farmed catfish and their environment (DePaola et al., 1988). The development of antibiotic-resistant pathogens in aquaculture products is certainly of concern to the consumer. The expansion of aquaculture in the future will warrant research on the development and better management of antimicrobials for aquaculture (Ward, 1989).

CONCLUSIONS AND RECOMMENDATIONS

Seafoods, like any food items, have the potential to cause disease from viral, bacterial, and parasitic microorganisms under certain circumstances. These disease-causing agents are acquired from three sources:

(1) fecal pollution of the aquatic environment;

(2) the natural aquatic environment; and

(3) industry, retail, restaurant, or home processing and preparation.

Fecal pollution may contribute human viral and bacterial contaminants and is the primary source of infection. Microorganisms associated with the natural environment include bacterial pathogens of marine origin and parasites transmitted from seafood to man.

Agents associated with workers and the environment in processing, distribution, food services, and home preparation include microorganisms carried by humans, as well as environmental microorganisms that become problems because of processing conditions.

Available CDC and NETSU data and literature reports suggest the following risk priorities for microbiological hazards in seafoods.

CONCLUSIONS

Raw Molluscan Shellfish

Overall, when examining the potential for seafood-associated illness from microbial pathogens, several factors must be taken into consideration, including host risk factors; sources and types of microorganisms; and seafood processing, preparation, and handling procedures that either allow microorganisms to survive and grow or destroy them before consumption. Food handlers and consumers must be made aware of all these factors. Imported products may have different levels of risk. Careful surveillance is necessary to monitor these risks adequately.

1. The greatest numbers of seafood-associated illnesses are reported from unknown etiologies clinically suggestive of Norwalk and Norwalk-like agents of human enteric viral gastroenteritis. The vast majority of these illnesses are associated with the consumption of raw molluscan shellfish taken from harvest waters contaminated with raw or poorly treated human sewage. Although these are the most common seafood-associated illnesses, they tend to be relatively mild with no associated mortality.
2. Naturally occurring marine *Vibrio* species are responsible for fewer reported cases of infections from the consumption of raw molluscan shellfish, but certain species such as *V. vulnificus* can be associated with high mortality in persons who are immunocompromised or who have underlying liver disease.

Other Seafoods

1. The greatest microbiological risk associated with seafood other than raw molluscan shellfish appears to be recontamination or cross-contamination of cooked by raw product or contamination during preparation followed by time/temperature abuse. This occurs mainly at the food service (postprocessing) level. The number of total reported cases associated with finfish between 1978 and 1987 is much lower than reported for raw

molluscan shellfish in the same period. Available data show that *Vibrio parahaemolyticus* is responsible for the largest number of other seafood-associated cases of illness, followed by hepatitis A virus, *Salmonella* (nontyphoidal), *Shigella, Clostridium perfringens*, and *C. botulinum*, with HAV and *C. botulinum* being the most potentially serious of these pathogens. However, the fish-associated HAV infections reported during this time were attributed to two outbreaks due to contamination of prepared seafood by infected food handlers. Seafood-associated illnesses due to *C. botulinum* were confined to a small geographical area (mainly Alaska) and were associated with the consumption of improperly processed noncommercial products.

2. Seafood-related parasitic infections are even less common than bacterial or viral infections, with *Anisakis simplex* and cestodes having the greatest public health impact in the United States. In general, parasitic infections are concentrated in certain ethnic groups that favor consumption of raw or partially cooked seafood harvested from high-risk geographic areas.

RECOMMENDATIONS

Specific Recommendations for Raw Molluscan Shellfish

High-risk groups (cirrhotics, persons with hemochromatosis, persons who are immunosuppressed) must not eat raw shellfish. It is extremely important that health professionals, especially, be educated concerning food-borne hazards to this group. Proper and thorough cooking of all shellfish before consumption would eliminate microbiological pathogens and helminthic parasites. Individuals who choose to consume raw shellfish should be educated about the potential risks described previously, and how those risks or their effects can be mitigated.

Adequate and proper treatment and disposal of sewage must be implemented and maintained to avoid contamination of harvest areas by human enteric pathogens. This may require the development of new technology for sewage treatment.

Valid indicators for contamination of growing waters by human pathogens must be developed. Seafood-borne infections by human enteric viruses in raw and improperly cooked molluscan shellfish could be decreased significantly by the development of valid growing water indicator(s) or direct detection methodologies for human enteric viruses. Effective enforcement for elimination of recreational and illegal ("bootlegged") harvesting or sale of molluscan shellfish from known sewage-contaminated shellfish growing areas should be developed and adequately funded. Monitoring programs for *Vibrio* species in molluscan shellfish and growing waters during warm months, as well as support for epidemiological research, should be established. Means must be investigated and implemented to eliminate, or at least reduce, levels of potentially pathogenic *Vibrio* species in raw shellfish. This may necessitate restriction of harvest when water temperatures are high, rapid cool-down and continued chilling of products, and possibly irradiation of live shellstock and shucked products. Because of the high risks associated with raw molluscan shellfish, the importation of shellfish for raw consumption should be prohibited unless there is a clear equivalence of standards for harvest waters and for postharvest processing.

General Recommendations for All Seafoods

Persons consuming raw fish or shellfish should be made aware of the potential microbial risks associated with these practices. Persons in specific high-risk groups (persons with cirrhosis, or hemochromatosis, or those who are immunosuppressed) should never eat raw seafood. Proper and thorough cooking of all seafood before consumption would eliminate the microbiological pathogens and helminthic parasites. Individuals who choose to eat raw seafood should be educated about the potential risks described previously, and how those risks or their effects can be mitigated. Any seafood inspection system must be designed to address microbiological hazards through the HACCP approach. This cannot be achieved by the visual or organoleptic inspection currently used for meat and poultry. Seafood inspection requires the development of valid microbiological guidelines to accurately assess potential human health risks from microbial pathogens in raw and processed seafoods; the maintenance of adequate refrigeration; the avoidance of recontamination of cooked, ready-to-eat products by raw products; and good manufacturing practices and proper sanitation. All inspection system guidelines must apply to imported as well as domestic products under memoranda of understanding.

More research is required to develop new technology-based processing and preservation techniques that provide for safe products. Some of the new processing methods such as *sous vide* and modified atmosphere can potentially create conditions that favor *Clostridium botulinum* type E and other pathogens such as *Listeria*. New methodologies that produce organoleptically superior products must also ensure superior microbiological safety. Certification procedures should be developed for any new processing techniques. Continuous, enhanced efforts should be undertaken to educate all health professionals, food handlers, and consumers regarding the microbiological risks of seafood-borne illness and the appropriate means of minimizing such risks, including immediate and adequate refrigeration, proper cooking, avoiding recontamination of cooked products by raw products, proper sanitation, and good personal hygiene, especially at the food service level.

A food-borne illness surveillance system sufficient for risk identification and regulatory program planning and evaluation must be developed. This system should provide a comprehensive data base that will allow statistically valid assessments of disease incidence and food/behavior risk factors for all food-borne illnesses. It is extremely difficult to assess the relative safety of seafood products accurately, or to manage seafood-borne or other food-borne risks effectively, with the available data bases. New or improved methodology [e.g., enzyme-linked immunoabsorbent assay (ELISA), gene probe, polymerase chain reaction] should be developed that provide for rapid identification and quantification of indicators, seafood-associated pathogens, and microbial toxins in seafoods and in harvest waters.

Bibliography

Ausubel, F.M.: *Current Protocols in Molecular Biology*, New York, John Wiley and Sons, 1989.

Barnes, N.: *Biology*, New York, Worth Publishers, 1989.

Barnum, Susan R.: *Biotechnology: An Introduction*, Belmont, Thomson/Brooks/Cole, 2005.

Brandwein, P.F.: *Sourcebook for the Biological Sciences*, San Diego, Harcourt Brace JOvanovich, 1986.

Brooks, Rodney A.: *Flesh and Machines: How Robots Will Change Us*, New York, Pantheon Books, 2002.

Clark, J.M.: *Experimental Biochemistry*, New York, W.H. Freeman and Company, 1977.

Curtis, H.: *Biology*, New York, Worth Publishers, 1989.

Dixon, Dougal: *After Man-A Zoology of the Future*, New York, St. Martin's Press, 1981.

Drexler, K. Eric: *Engines of Creation: The Coming Era of Nanotechnology*, New York, Anchor Books, 1986.

Friedberg, E.C.: *DNA Repair*, New York, WH Freeman and Company, 1985.

Gaskell G: *Biotechnology-the Making of a Global Controversy*, Cambridge, Cambridge University Press, 2002.

George S. Paul: *Beyond Humanity: Cyber Evolution and Future Minds*, Roackland, Charles River Media, 1996.

Goodman, D.C.: *From Farming to Biotechnology: A Theory of Agro-industrial Development*, Oxford, Blackwell, 1987.

Goodsell, David S.: *Bionanotechnology: Lessons From Nature*, Hoboken, Wiley-Liss, 2004.

King, R.B., Mihony, G., and Sheldon, J.K.: *Practical Environmental Bioremediation*, Lewis Publishers, Boca Raton, 1992.

Kurzweil, Ray: *The Age of Spiritual Machines*, New York, Penguin Books, 1999.

Moravec, Hans: *Mind Children: The Future of Robot and Human Intelligence*, Cambridge, Harvard University Press, 1988.

Morholt, E. : *Sourcebook for the Biological Sciences*, San Diego, Harcourt Brace JOvanovich, 1986.

Murray, David: *Seeds of Concern: The Genetic Manipulation of Plants*, Sydney, University of New South Wales, 2003.

Old, R.W. : *Principles of Gene Manipulation*, London, Blackwell Scientific Publications, 1989.

Parens, Erik: *Enhancing Human Traits: Ethical and Social Implications*, Washington, D.C.: Georgetown University Press, 1998.

Postman, Neil: *Technopoly: The Surrender of Culture to Technology*, New York, Vintage Books, 1992.

Primrose, S.B.: *Principles of Gene Manipulation*, London, Blackwell Scientific Publications, 1989.

Retzer, W.J. *Biotechnology Workbook*, Englewood Cliffs, Prentice Hall, 1991.

Rifkin, Jeremy: *The Biotech Century*, New York, Penguin Putnam, 1998

Sayler, G.S., Fox, R. and Blackburn, J.W.: *Environmental Biotechnology for Waste Treatment*, Plenum Press, New York, 1991.

Shetty, Kalidas: *Food Biotechnology*, New York, Dekker/CRC Press, 2005.

Slater, R.J.: *Experiments in Molecular Biology*, Clifton, Humana Press, 1986.

Smith, John E.: *Biotechnology*, Cambridge, Cambridge University Press, 2004.

Switzer, R.L.: *Experimental Biochemistry*, New York, W.H. Freeman and Company, 1977.

Towle, Albert: *Modern Biology*, Austin, Holt, Rinehart and Winston, 1988.

Walden, Richard: *Genetic Transformation in Plants*, England, Open University Press, 1988.

Index

A

Aquaculture, 1, 5, 17, 18, 19, 21, 25, 117, 118, 126, 130, 134, 135, 140, 144, 152, 195, 198, 199, 200, 201, 205, 214, 271, 272.

Aquatic Genetic Resources, 130, 131, 133, 134, 135, 140, 141, 142, 144, 145, 147, 151, 152, 156, 158, 161.

B

Biochemistry, 11.

Biotechnology, 145.

C

Chemical Contaminants, 1, 4, 24, 30, 45, 113, 114, 126, 130, 132, 135, 208.

Chemical Residues, 2, 3, 115, 116, 126.

Community, 66, 102, 104, 123, 135, 136, 141, 142, 146, 147, 149, 150, 151, 153, 157, 158, 159, 161, 172, 204, 218.

Culture, 17, 18, 20, 201, 219, 220, 245.

D

Data Collection, 121, 128.

Development, 20, 33, 35, 47, 48, 50, 59, 69, 72, 106, 111, 114, 115, 120, 121, 122, 123, 125, 126, 133, 140, 144, 148, 151, 152, 154, 157, 167, 174, 175, 187, 189, 190, 193, 208, 261, 269, 271, 272, 274, 275.

E

Evidence, 5, 6, 8, 13, 42, 50, 57, 72, 163, 172, 174, 181, 216, 229, 258, 270.

F

Fish Culture, 17, 20.

Freshwater Environments, 119, 215, 259.

H

Health Risk Assessment, 3, 24, 26, 255.

I

Implications, 2, 26, 37, 56, 63, 134, 160, 161, 206.

Improvement, 45, 125.

Industry, 9, 19, 22, 52, 108, 114, 118, 120, 122, 126, 127, 128, 135, 144, 145, 165, 180, 190, 191, 192, 195, 196, 197, 199, 200, 202, 203, 207, 253, 263, 266, 269, 271, 272.

Inland Waters, 102, 132, 133.

Institutions, 3, 153, 158.

L

Laboratory, 57, 75, 95, 96, 97, 240, 242, 250.
Legislative Considerations, 126.

M

Management, 1, 2, 3, 18, 25, 117, 130, 131, 133, 134, 138, 158, 159, 160, 190, 213, 227, 243, 244, 259, 272.
Marine Fish, 65.
Microorganisms, 19, 108, 110, 118, 124, 129, 165, 183, 192, 197, 209, 213, 214, 217, 238, 253, 259, 267, 268, 271, 272, 273.

N

National Sovereignty, 132, 133, 134, 135, 138, 139.
Natural Toxins, 107, 112, 114, 115, 118, 121, 124, 129, 162, 164, 165, 197.

O

Opportunity, 55, 57, 115, 162, 203.
Organization, 18, 74, 83, 95, 145, 146, 148.

P

Parasites, 17, 107, 108, 112, 197, 209, 232, 261, 270, 272, 274, 275.
Prevention, 51, 113, 167, 176, 190, 226, 238, 246, 261.
Project, 132, 160, 245.
Property law, 137, 141, 142, 149, 150.
Public Health Risks, 106, 123, 126, 162.

R

Relationships, 26, 27, 30, 33, 37, 42, 46, 57, 60, 61, 205.
Religious, 135.
Research, 18, 24, 25, 26, 36, 43, 63, 72, 80, 83, 84, 90, 95, 103, 109, 110, 117, 128, 129, 131, 132, 133, 134, 139, 144, 147, 152, 153, 154, 155, 158, 159, 160, 174, 175, 183, 190, 191, 192, 193, 194, 208, 220, 266, 267, 272, 275, 276.
Revolution, 66, 133.
Risk Management, 1, 3, 213, 243, 244, 259.

S

Seafood Industry, 19, 126, 197, 200, 253.
Shellfish Poisons, 166.

T

Technology, 22, 132, 139, 142, 144, 145, 205, 208, 252, 254, 260, 267, 269, 274, 276.
Traditional Communities, 134, 135, 136, 142, 151, 152.
Treatment, 1, 18, 21, 22, 46, 47, 59, 63, 64, 109, 111, 113, 115, 138, 150, 158, 177, 188, 192, 208, 218, 237, 238, 247, 252, 253, 274.

♦♦♦